THE OIL MAKERS

THE

OIL

MAKERS

INSIDERS LOOK AT THE PETROLEUM INDUSTRY

Edited by Jeffrey Share

Introduction by Joseph A. Pratt

RICE UNIVERSITY PRESS/HOUSTON TEXAS

Requests for permission to reproduce material from this work
should be addressed to

Rice University Press
Post Office Box 1892
Houston, Texas 77251

Book design by Patricia D. Crowder

LIBRARY OF CONGRESS CATALOGING-IN-PUBLICATION DATA
Share, Jeffrey, 1951–
The oil makers / edited by Jeffrey Share.—1st ed.
p. cm.
ISBN 0-89263-339-5 (hc).
1. Petroleum industry and trade—United States—Forecasting.   2. Gas industry—
United States—Forecasting.   3. Petroleum industry and trade—United States—
Employees—Interviews.   4. Gas industry—United States—Employees—
Interviews.   I. Pratt, Joseph A.   II. Title.
HD9565.S438 1995
338.2'8282'0973—dc20   95-33943
                                        CIP

This book is dedicated to the men and women of the American petroleum industry.

ACKNOWLEDGMENTS

No undertaking such as this book can be a one-person effort. I am extremely grateful to the hundreds of people who assisted me in one way or another. I first want to thank my former coworker at the now-defunct *Houston Post*, Dave Baranowski, for motivating me to undertake this project, and another former coworker and close friend, Jack Boone, for his support and unlimited patience nearly every day that it took to complete *The Oil Makers*. I'm also thankful to Susan Bielstein, the editor-in-chief of Rice University Press, who believed in this project from the moment we spoke. Dr. Joseph Pratt from the University of Houston helped shape the final product, lent his name, and also became a good friend.

Among those who deserve a very special thanks are Al DeCrane, William Stevens, Lodwrick Cook, Richard Stegemeir, C. J. Silas, Henry Groppe, Jerry Brown, John Bookout, Robert O. Anderson, Kenneth Jamieson, John Swearingen, Edward Price, William Greehey, Michael Beatty, Dennis Hendrix, Russ Luigs, Bob Palmer, Tom Bowersox, Jim Nelson, Michel T. Halbouty, Pat Rutherford, James Payne, Robert Casey, Dale Steffes, Al Wadsworth, Forrest Hoglund, Paul Howell, Doug Rock, Gene Ames, George Alcorn, James D. Woods, Ike Kerridge, Roy Caldwell, Ray Plank, Robert Mosbacher Sr., Jack Bowen, Boone Pickens, Victor Burk, James Crump, Charles DiBona, Charles Blackburn, William Hochheiser, John Lollar, Lynn Leigh, W. Henson Moore, Congressman Tom DeLay, and Robert Scott.

I also want to thank Mike Dixon, Sondra Fowler, Kitty Borah, Rob Phillips, Mickey Driver, Bobbie Bounds, Robert Bradley, Diane Baziledes, Debra Vondra, Becky Virtue, Susan Nardone, Kate

Hutcheons, Patrick Cassidy, Judy Ames, Cynthia McMahon, Robert Harper, Anna Perraut, Shelly Daughtry, Suzanne Darnell, Macy Moy, Dan Ammerman, Jim Hart, John Barnett, Brandon Lackey, Linda Dolan, Linda Silinsky, Tom K. Stewart, Bill Gilmer, Stephen P. A. Brown, Dr. William Fisher, James Regg, Rick Stetter, Josh Galper, Miles Glaspy, and Les Mallory.

Thanks to Barbara Boone, who did much of the transcribing, and Kimberly Trainer Carroll, who also patiently sat through endless hours of transcribing and helped name *The Oil Makers*.

In looking back over all of the people in and out of the industry who assisted me during the three years it took to complete this book, I don't think I can ever express my full gratitude.

Finally, I assume all responsibility for accuracy of the text, including biographical introductions and footnotes.

—JEFFREY SHARE
HOUSTON
1995

Contents

THE OIL MAKERS

INTRODUCTION

Voices of Change

We are living in an era of sweeping change in the energy industries. Yet few Americans outside the oil, natural gas, and electric industries are aware of the pace or direction of this change. The popular perception is that oil and gas are declining, moving steadily toward a day when domestic production will be a minor economic activity, an insignificant sideshow in the circus of international oil. This view ignores clear signs of a far-reaching transformation in the ways we find, produce, market, and use energy in America. For much of the twentieth century the oil, gas, coal, and electric industries have managed our energy supplies, with each largely separate from the others and all heavily regulated by various levels of government. A more unified, less regulated, and more market-driven energy industry is emerging in the 1990s.

Public discourse about energy seems trapped in the past, with too little reference to the ongoing transformation of these vital industries. Traditional ways of thinking about energy hold sway; public policy debates remain focused on issues inherited from the 1960s and 1970s. This lag in perceptions reflects, in part, the natural workings of a democracy, which requires time and often a crisis to mold a new consensus on controversial issues. Critics of the industries have set the public agenda of debate in recent decades, with industry representatives reacting to issues raised by opponents rather than putting forth a systematic view of the accomplishments and needs of their industries. Although many lament the public's ignorance of their industry, few have found effective ways to educate the public.

Thus, as competitive forces raise critical questions about energy supply and demand, the public hears few convincing answers from within industry. Fundamental divisions within the energy industries have produced a Tower of Babel, as a confusing array of voices from competing interests yields a jumbled message that few can understand. Domestic oil companies have markedly different self-interests than the international majors; oil, gas, coal, and electricity often vie for capital, markets, and political influence. Inevitably, competing voices have been heard from these industries on every important issue. Leaders within each industry generally lack either the experience or the incentive to look beyond such divisions and offer a convincing vision of what these industries have in common and how they fit together to satisfy the nation's expanding demands for energy.

Changes in markets and technology are beginning to force the discussion of such issues to the fore. Computers have given us a much greater capacity to manage the flow of all forms of energy from production to market, encouraging the comparison of price and availability over larger and larger geographical areas. At the same time, the loosening of government regulation has encouraged innovators and entrepreneurs to compete more aggressively for new and existing markets for energy. As traditional boundaries between the various energy industries blur, the public—in the form of investors, workers, consumers, and voters—will need a clearer understanding of the movement toward a more competitive, more unified, and more international market for energy in all of its varied forms. Who will provide such understanding?

One logical place to look for guidance is among the leaders of the oil and gas industries, which are central to our energy future. As an energy writer, Jeffrey Share had interviewed many of these leaders, and he felt that their views deserved a broader audience. He set out to collect and publish a series of interviews that would communicate the views of a selected group of industry leaders. He talked with representatives from international and domestic oil companies, from the oil and gas service sectors, from financial institutions with ties to the energy industries, from various levels of government, and from the natural gas industry. Some of those interviewed are long-established industry leaders who sit at the head of major companies; others are up-and-coming leaders from particularly dynamic sectors of the energy-related complex of industries. All have years of experience that give them insiders' perspectives on key issues.

The question Share posed at the beginning of each interview was straightforward and to the point: What do you consider the key issues facing the energy industry? He allowed those being interviewed to speak for themselves, framing their own answers with a minimum of prodding to address specific issues. The taped interviews generally lasted an hour or two; transcriptions were returned to the interviewees, who were then encouraged to make any corrections or additions they wanted. After completing a preliminary round of almost sixty interviews, he conducted follow-up interviews with many of the thirty people selected for inclusion in this book. The completed interviews flow in many different directions, reflecting the perspective and position of each individual. Brief biographical sketches of the persons interviewed are included to give the reader a sense of the experiences that have shaped their views. These interviews show clearly the diversity of views within these industries, but they also suggest several general trends and concerns common to much of the industry.

The first is a widely shared sense that the energy industries are on the verge of a new order after a long and often painful era of transition. The OPEC-induced oil price shocks of the 1970s destroyed the relatively stable era of predictable energy prices that had marked the post-World War II order in oil and gas. For almost thirty years after the war, the oil and gas industries enjoyed steady expansion under the predictable, if at times constraining, regulatory rules of the Texas Railroad Commission in oil and the Federal Power Commission in natural gas. This era of steady expansion ended abruptly in the 1970s, as OPEC imposed staggering increases in oil prices that fed a chaotic boom in the 1970s. A stunning bust followed in the 1980s, as prices plummeted with the loosening of OPEC's control. (See Chart 1.) This largely unpredicted freefall devastated much of the domestic oil industry, forcing painful adjustments in strategies based on the false expectation of steadily rising oil prices. Whereas numerous experts in the 1970s forecast the coming of a new OPEC-dominated order, the 1980s brought little discernable order of any sort, as OPEC proved vulnerable to internal tensions, conservation, and the production of large quantities of non-OPEC oil. Thus for more than two decades the petroleum industry faced considerable uncertainty over both the price and the supply of crude oil. The cycle of boom and bust in oil spilled over into related energy industries, fostering an era of anxiety.

Finally, in the 1990s, many industry leaders recognize the outlines of something resembling a new order in oil, one characterized by relative price stability and predictable sources of supply, at least over the short to medium run. The new order in the U.S. includes reliance on imports for as much as half of the nation's oil supply, with a measure of security from disruptions in supply coming from the diversity of suppliers, the growth of non-OPEC imports, and the use of public policies ranging from the Strategic Petroleum Reserve to diplomatic and military initiatives. (See Chart 2.) With the prevailing international price for crude oil too low to sustain the expansion of domestic oil production, this new order promises continued hard times for oil exploration and production in the U.S. In a much-drilled "mature" oil province such as the United States, even efficient and innovative oil companies face an uphill battle in competition with international companies active in the Middle East and other producing regions with larger, newer, and lower-cost oil fields. This explains one recurring tension obvious in the interviews, the debate over the need for some sort of public policy to bolster the domestic oil industry. Those dependent primarily on domestic production—from drillers to producers to oil tool manufacturers to congressional representatives—have a strong incentive to look to government to provide relief for the domestic oil industry from the ruthless devastation brought by international market forces. The debate over the need for such relief and the form it should take remains a divisive issue in American energy politics, and no other set of issues so clearly illustrates the diversity of viewpoints within the oil industry.

A more fundamental difference in viewpoint is evident in the interviews with executives in the natural gas industry. Long considered the junior partner in the oil and gas industry, natural gas is becoming an increasingly important source of supply in what may be referred to as the domestic gas and oil industry. While prospects for domestic oil production remain bleak, the prospects for gas appear much brighter. Before

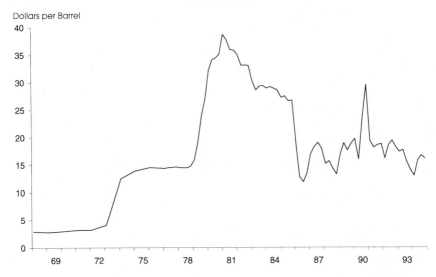

World Oil Price

Chart 1

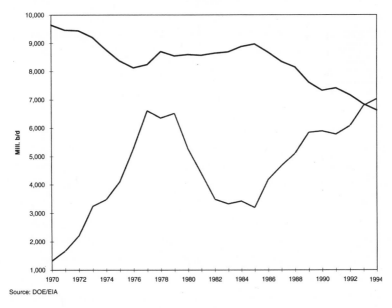

Domestic Crude Oil Production & Imports

Source: DOE/EIA

Chart 2

phased deregulation in 1978, natural gas was a comfortably regulated industry with little competition to encourage innovation. Phased deregulation brought an extended period of uncertainty, as the Federal Energy Regulatory Commission (FERC) gradually allowed the introduction of new competitive pressures; now in the mid 1990s, market-driven enterprises dominate the reorganized natural gas industry. Interviews with strategists in this dynamic industry suggest strongly that the competitive impulses unleashed by deregulation might spill over the traditional boundaries of natural gas, becoming a source of creative change in such related sectors as electricity and even oil.

Such dynamism is not limited to natural gas; indeed, interview after interview points out significant examples of far-reaching changes at work in the energy industries. The economic pressures that brought hard times to the industry also brought an overwhelming incentive for many companies to innovate or die. The most visible and highly publicized response has been often severe cutbacks in employment, as companies respond to economic imperatives to become "leaner and meaner." Severe personal hardships and much bad publicity have accompanied massive layoffs in recent years. Industry spokespersons tend to exaggerate the magnitude of the cutbacks in oil by using the highly unusual temporary boom years of the early 1980s as the baseline for measuring subsequent employment losses; but by any measure, the last decade has witnessed far-reaching changes as much of the oil-related economy has been forced to reorganize. (See Chart 3.) Despite the obvious personal costs of this process, more efficient and flexible organizations are emerging in much of the oil industry.

In other areas, the dynamism of the industry lies hidden from public view. Numerous interviews point out dramatic technological innovations that are enhancing the capacity to produce and distribute energy. Ongoing advances in "hardware" technology continue a long tradition in the energy industries, which have excellent historical records of developing the equipment needed to produce ever-greater supplies of energy. Such innovations as directional drilling and 3-D seismic processes for finding oil represent extraordinary advances of great value in expanding oil production. More fundamental change is occurring in the form of "software" technology. Several of the interviews offer fascinat-

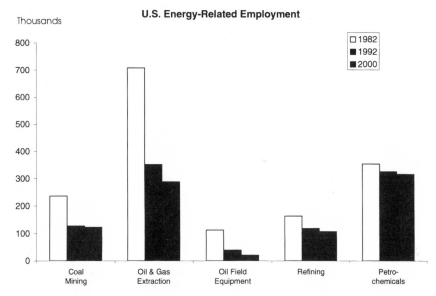

Thousands

U.S. Energy-Related Employment

Chart 3

ing speculation about the ultimate impact of computers in enhancing both consumer choice and industry efficiency. Computer technology has the potential to create ever-larger markets across both national boundaries and the traditional boundaries between the different suppliers of energy. The somewhat awkward phrase "commoditization of energy" now in use suggests the possible long-term impact of the computer revolution in energy; in the future, consumers might be able to choose almost instantaneously among different energy options on the basis of current information about price and availability. Continued improvement in the capacity of computers to track the delivery of energy within a national and international grid pushes us toward a future in which energy might become more and more like other commodities—with the traditionally separate industries of oil, gas, coal, and even electricity gradually merging into a unified market for energy in a variety of directly competing forms.

The starting point for such future changes has been the coming of more competitive markets for oil and gas. Although OPEC remains a significant power in international oil, its inability to impose effective controls on the supply or the price of oil has opened the way for greater price competition. Deregulation of oil prices in 1979 and 1980 and of natural gas prices after 1978 also encouraged greater competition within the domestic energy industries. The political implications of this movement toward greater reliance on the market in energy are touched on throughout the interviews, which return again and again to modern variants of questions as old as John D. Rockefeller: To what extent do the implications of a healthy domestic oil and natural gas industry for national security override strictly economic interests? To what extent will Americans be content to allow power over energy to be concentrated in the hands of private businesses? What sort of government regulations are needed to safeguard the public's interest in this vital sector of the economy?

When it comes to politics, one issue looms larger than others in the interviews. The most controversial public issue facing the energy industries remains the trade-offs between energy and environment. Many of the interviews raise a litany of complaints against environmental regulators, which, by now, has become commonplace: the high costs of compliance with government regulations, the inflexibility of standards, and the lack of sensitivity to the needs of the oil industry—particularly in prohibitions against drilling on government lands. Taken as a whole, such complaints remain the most significant source of conflict between the oil makers and the broader public. There are some signs of change in this controversial area, most notably proposals for regulatory reforms that advocate the use of economic incentives to encourage cleaner operations. But most of the interviews suggest that a more efficient and more sustainable system of environmental regulations is a much needed—and as yet missing—element of a sustainable new order in energy. Repeated assurances that oil people are environmentalists at heart have not resolved the fundamental philosophical and political differences between those who have committed their lives to the production of energy and those who have made equally strong commitments to working for a cleaner environment. Legal and political conflict no doubt will continue over these issues, as the nation seeks to define an appropriate energy/environmental policy.

These and many other key issues raised in the interviews ultimately concern all of us who live in an energy-intensive society. Whatever our personal stance on such issues, we should listen seriously to the views of those most closely involved in the energy business. In the following interviews leaders of the industry speak for themselves. We invite the reader to see the faces and hear the voices of these individuals, whose long years of experience yield valuable views from inside the energy industries.

JOSEPH A. PRATT
UNIVERSITY OF HOUSTON
SUMMER 1995

Overview

ROY M. HUFFINGTON

chairman of Roy M. Huffington Inc.

Roy M. Huffington's life story is a testament to what a man can achieve by combining talent, vision, and persistence with a willingness to take risks. A touch of gambler's luck doesn't hurt, either. Huffington ultimately succeeded beyond his own hopes or imagination, becoming one of the twentieth century's legendary international explorers of petroleum.

The son of an oilman who was killed in a platform accident in Venezuela in 1932, Huffington was born near Houston and received a B.S. degree from Southern Methodist University and M.A. and Ph.D. degrees in geology from Harvard University. Decorated twice while serving in the Navy in World War II, he joined Humble Oil & Refining Company (the forerunner of Exxon) in 1946 as a field geologist, leaving in 1956 to set up his own company. In 1968, after the Federal Power Commission ruled that the price of his company's gas was too high and ordered him to pay back his customers, Huffington turned his focus overseas and secured a production-sharing agreement with Indonesia. He took a gamble that no major would have taken, and, ultimately, his discoveries in the jungles of East Kalimantan led to the development of a multi-billion-dollar LNG export project between Japan and Indonesia, where he is regarded as a national hero.

An entrepreneur who has worked on projects in sixty countries, Huffington was chosen by President George Bush to serve as ambassador to Austria in 1990. After selling the overseas properties of his company to the Chinese Petroleum Corporation of Taiwan, he served his country with distinction, helping to open up business opportunities between newly independent Eastern bloc countries and the West. He left the ambassadorship in 1993 to return to Houston, where he is again the head of Roy M. Huffington Inc., an independent international oil and gas company.

Between flights around the world, Huffington made time for a conversation in his suite on the sixty-seventh floor of a downtown Houston skyscraper. He is a courtly man with a winning smile and a firm handshake guaranteed to put anyone at ease. Many insist, rightly so, that his voice resembles that of John Wayne, and Huffington's real-life exploits rival that actor's on-screen adventures.

The *real key* problem that the oil and gas industry has, particularly in the United States, is the lack of true understanding by politicians of what the oil and gas industry is all about. In the United States, they have removed many areas from exploration, such as the continental shelf off the west coast of California. As a result, there are many areas that people can't really explore, both onshore and offshore in the United States. I think that will change, after U.S. oil and gas get to be shorter in supply. Currently, we have to import over 50 percent of our oil and gas requirements in the United States.

I don't think the oil and gas industry will ever come back the way it was, say in the late '70s and early '80s. But there is still a tremendous amount of oil and gas that we could find in United States territory, if we didn't have so many political and environmental restrictions.

The oil and gas industry can work almost any place in the environment without polluting it. I think that the politicians need to recognize that, when we have undeveloped reserves waiting to be discovered, they should let the industry discover these reserves—for the time being, at any rate. When we begin to run out of oil and gas in the States, we will still have tremendous coal reserves to develop. The use of coal is going to be a little more polluting, and a lot of work will have to be done to scrub and clean up coal-burning power plants and other places where coal may be burned.

Overseas, also, if it weren't for the politicians in many countries, I believe we could literally cover the Earth with a foot or two of oil extracted from the many places where one might explore. But so many times, one can't make a fair or reasonable deal in a given country and, as a result, that country stays relatively undeveloped.

In a little talk that I made the other day, I mentioned just briefly that some politicians have their own, and differing, agendas that restrict the development of natural resources. Then, in some countries, there is also a great deal of corruption. In those countries, it is impossible for American companies, with our current laws, to work there. We are not going to attempt to bribe anyone, which sometimes other groups might be willing to do. But, in my opinion, the political factor is still the largest handicap for continued development of the oil and gas industry in most countries.

Realistic regulations

Now, environmental regulations are pretty stiff in a lot of places. When we first went to Indonesia, we actually helped set up a little group there to protect the environment, to be sure that people didn't destroy the jungle—whether it be logging groups, or oil and gas exploration groups, or anyone else—that they didn't leave the land in such bad shape that it would take some time for nature to correct what had been spilled on it, or what had torn it up.

But I know that in some areas, the so-called environmentalists go to such extreme lengths that all of the rules and regulations they set up really do not make any common sense. I noticed the other day in Texas that some regulatory group had passed a restriction in the drilling of a new well that would prevent you from disposing of salt water in certain ways. Thus, you might have next to the new well an old one from which you are producing some salt water and reinjecting it into an adjacent disposal well, but yet you can't use that reinjection well for salt water for the new well.

These are confusing regulations that originate in big bureaucracies, where a group at one end of the organization doesn't know what a group at the other end is doing. Thus, environmental problems are a little more restrictive for the industry than they should be. The oil industry is quite capable of keeping the air, land, and water pristine clean.

I think most oil companies, both large and small, are environmentally knowledgeable enough that they are not going to pollute anything. There may be some trade-offs, or changes, that currently need to be made, where environmental laws have gotten so restrictive that they are driving the cost of exploration much higher than it should be.

But I don't think anyone whom I know in the industry is upset about having normal environmental requirements. They are upset mainly about the complicated nature of some of the laws, so that if you obey one, you may not be able to obey another. So many bureaucratic rules have been written that it's very difficult to comply in some places. I know, for instance, that one of the laws passed some time back, for operations out in the Gulf of Mexico, made every operator carry about $150 million worth of insurance, in case anything went wrong.

Well, that just takes exploration out of the hands of the independents, who are doing 85 percent of the work out there, because they can't pay for that amount of insurance. If you look at the record, there's been very little pollution in the Gulf of Mexico.

Now, I'm a lover of the environment, the clean air, the clean water, and all of the things that I seem to recall when I was growing up. A clean environment is something that all mankind wants and should continue to have. The oil and gas industry, itself, is quite capable of protecting the air and water, and I think it has done so in general, with the exception of a few limited and generally accidental cases.

The other thing that people really need to think about is that oil is a biodegradable substance. It's messy as can be, if you have a spill. There isn't any doubt about that. I would hate to be around where some oil had been spilled, because it's gummy. It sticks on your shoes or feet. It messes up the beach. But a year later, it has usually all disappeared. For millions of years, nature has been spilling oil out into the oceans and onto the land, where there have been seeps, and it hasn't destroyed anything. Little microbes start eating the oil, and they finally destroy it, if it's in small quantities.

Evolution of technology

Technologies will continue to evolve and improve. In the last two decades or so, it's amazing how much various techniques have improved. Geophysical work being done in some difficult areas of the world is now getting results in mapping structures at depths where it was impossible to get good records twenty or thirty years ago. Now geophysicists can. So-called 3-D seismic exploration is outlining in detail little fault segments on given structures, which has enabled explorationists to find additional reserves around areas that have already produced some.

Our technology is soon going to be good enough to allow us to drill wells in 10,000 feet or more of water, but the price of oil and gas has got to be a lot higher than it is now to enable the industry to produce such wells economically. We can certainly drill and find oil and gas in those areas, but it's going to be very expensive to produce it.

Groundless fears

There are still a few other problems that the industry has, but these would probably fall in the nature of problems that any business has as it continues to expand and develop. In the oil and gas industry, we are particularly concerned about restrictions that keep us from drilling in many places around the world. This is particularly true in the U.S., where a lot of territory has been restricted to drilling by the federal government.

I think people are opposed to drilling in some areas, basically because they don't like something similar to construction work going on near their house. Similarly, when you start to widen streets or build a freeway, it takes a long period of time, and people get irritated. It's inconvenient. I think that's the main reason people don't like it.

Anyone who knows anything about the oil business is usually sort of excited if there's some drilling in the area. Maybe the operator might later move over and drill on a piece of land that you own. If they find some oil under your land, that is always tremendously exciting, because it means you will have additional income. I think the main reason people object to drilling is the inconvenience of things, or sometimes because a drilling rig doesn't look pretty, or makes noise while you are drilling.

Now offshore, particularly in California, people just don't like anything that's unusual. Yet on some of the drilling islands that they have put off the coast of California, they have planted a few palm trees and shrubs, and they look like true little islands off the coast. It doesn't look bad at all. I always thought it was sort of nice, at night, to see a few lights out there and realize that, out in the middle of the ocean, you have a small community on a little island where men are working to bring oil and gas ashore. This activity is to help the people of our country by supplying fuel to drive their machinery and to drive their cars.

But at the same time, a few environmentalists have gone too far in their vision. Some would like to go back, I believe, to the horse and buggy days. However, anyone that has ever been around horses and buggies in years past much prefers the cleaner car than they do the old horse and buggy.

Level playing field

Government needs to be sure that businesses are competing fairly— that you don't have complete monopolies, where someone gets total control of a particular business. I think they need to be sure that there is competition in all businesses, not just oil and gas. They need to be sure, of course, that businesses pay whatever their fair share of taxes might be, according to the tax laws in the land at that time.

I think government also needs to quit saddling businesses with so many convoluted, detailed laws and regulations that are almost impossible to understand. For instance, just like the income tax laws. I don't know how many thousand pages of regulations actually exist. Every time they change the tax laws in Congress, even the tax accountants and tax lawyers sometimes have a hard time understanding what is actually proper. It's pretty hard to get the Internal Revenue Service to give you a letter or ruling on anything, because they'd rather look at you after the fact than tell you beforehand that this is the way you have to do it.

Many governmental laws and regulations are too complex and bad. In my case, the old Federal Power Commission, in one sense, really ran me out of the country in about 1967, or thereabouts. They had given us a tentative gas price, when they were controlling gas prices, of say, 24 cents a thousand cubic feet. I think about seven years later, they said in the last administrative hearing before the judge in Washington, D.C., that they were sorry, but that wasn't the final price. It should have been 21.625 cents, and maybe, in one case, 18 cents. So we were to give back the over-collection for those perhaps six or seven years in between, plus, as I recall, pay perhaps 7 percent interest on the over-collection.

Well, that meant that, as an independent, I was basically broke, because no one kept that amount of money in the bank. You were working the money all the time. I also had partners in with me, of course, and some of those had died. It wasn't totally clear as to whether they owed some of the money or if we, as the operators, owed the money. We had to work out a deal whereby we actually made arrangements for them to recover that amount of money out of future production. With that, I decided to go overseas, because there was no way I could continue to build a company, if I never knew for how much I could sell my product.

The free market didn't reign. The government was controlling the markets, and yet they wouldn't give you a firm price, when they first decided. After that, they would run you through hearing after hearing after hearing. All those legal and travel costs amounted to a tremendous amount of money over several years, and you never knew exactly what you were finally going to get for your gas.

That's where I think government needs to stay out of the hair of business with all these complicated laws. That's one reason people are now beginning to push for either a flat tax or a sales tax or a value-added tax. I think it's a lot simpler than this complicated deal where most people don't understand what it's all about. I can't even tell you. I can't make out my own tax return. It's so complicated that I wouldn't know how to interpret all those thousands of pages of rules.

Life after oil

We'll become more gas-oriented for two reasons. Where we can use gas, it's even cleaner than oil. The other reason we'll be using more gas and less oil is because we're finding more gas now and less oil.

But oil, itself, will still be a very necessary thing for lubricants and for fuels, such as gasoline for cars. Oil is used in various synthetic products—rubber and other petrochemical things that have been developed by the petroleum industry. There will continue to be a real need for oil for primary fuels and power generation, although gas is much cleaner in power generation.

Ultimately, and I don't think there's any question about it, we'll get back to nuclear power for power generation, and that's where we really should be. A nuclear plant, properly constructed with today's technology and properly maintained, is no hazard to anyone. I wouldn't be afraid to live next door to one as long as it's properly maintained, because it's clean energy. The one thing, however, of which people constantly have to be aware is that you have to be very careful and maintain the plant properly every living, existing day. You can't say, "Well, it's running fine, so I don't have to worry about it today." You have to have safety-conscious people constantly checking all your controls to be sure that nothing is starting to go wrong accidentally. Or that some little metal part is beginning to fail.

The press, I think, exaggerates the issues and frightens many people who don't understand the technology and mechanics. "If this plant blew, it could destroy the state or half the country." Well, that's not going to happen. The plant at Chernobyl, in the former U.S.S.R., wasn't maintained as it should have been, and that has been the danger. They've done a lot better job now, and they've got outsiders also working with them to assure that the plants will hold up all right.

I don't care whether it's in the United States, Europe, or elsewhere around the world, most people are afraid of nuclear energy, and, politically, it's sort of devastating for a politician to recommend using it.

There is no pollution other than the radioactive remains, which need to be properly disposed. That can be done, by burial, in ways that won't pollute air, water, or anything else that will interfere with human habitation.

Some of the oil companies may get into the nuclear side. Exxon and various others, at one stage, had groups that were working on nuclear power. Then, because of all the rules and the confusion that surrounded government controls, most of them finally threw up their hands and said, "The heck with it. We'll forget about it." But ultimately, I think that petroleum companies should and will return to the nuclear power plant business.

But it's hard to get a permit to build one. When you get it, the government, in years past, has changed the rules and regulations. You start constructing one, and before you get too far along, they may have changed some of the requirements so you have to redesign a portion of it. As a result of that, practically every nuclear plant in the U.S. is designed a little bit differently, as far as details go. France, on the other hand, set up, I believe, 900-megawatt and 1300-megawatt plants with a standard design. They now get about 85 percent of their electrical energy or power out of those nuclear plants, and they're not having any problems. They are all basically designed the same way, so that if someone had to be called from one plant to go over to another one to fill in for a person that was ill, they would know already how to run it or what do to; they know what part they play, and that's not the case in the United States.

If it weren't so difficult and so expensive, when you get so many cost overruns because of the changing of laws while you're in the middle of the construction process, I think you could cut the cost of nuclear plants in half from what they have been in the past. In fact, I'm positive you can.

Lack of capital

Lack of adequate capital is currently a pretty sizeable problem, not only for independents, where it's always been a problem, but even for the major companies, because of escalating operating costs in both exploration and production. Major companies don't have all of the capital that they need to take on too many big projects around the world. I think that will continue to be the case, probably for an indefinite period of time.

We could spend hundreds and hundreds of billions of dollars developing the industry, and sometimes you could spend all of that in one country. Just look at Russia and the republics of the former U.S.S.R. I don't know how many hundreds of billions of dollars they could use just in developing some of their known fields and in rebuilding the infrastructure they need. There are tens of thousands of miles of pipeline needed to get the oil and gas out to market.

I think a lot of independents can still make a fair living in discovering oil and gas in certain areas, but it's going to be more limited than it's been in the past. Many areas have been rather densely drilled, at least to shallow depths. Some areas have had wells drilled every square mile or so. It's going to be more difficult to find reserves in these areas, although 3-D seismic work has helped a lot.

The biggest problem right now, as far as independents are concerned, is the lack of capital. It is very difficult to get any capital at this stage, because Congress changed the laws and basically removed the depletion and intangible drilling allowances that our forefathers, in their wisdom, previously established. In years gone by, we had a 27.5 percent depletion allowance, which meant that once you drilled a producing well, 27.5 percent of what you produced was tax-free. This was to give you an opportunity to recover your money sufficiently soon, so that you could drill again in a reasonable period of time. Let's say you spent a million dollars digging the well. After you spent your original million dollars, if you had to wait to recover it over a thirty-year period, you were out of business. It would take you thirty years to get your money back. With the 27.5 percent depletion allowance, in less than four years you had your money back, and you could soon be drilling again.

That's been the big problem, together with the handling of intangible drilling costs. All sorts of things that used to be expensible are, in general, not similarly expensible any more. Since most of the wells you drill as a wildcatter are dry, you go broke pretty fast if you're not lucky on the first well or two.

I think there are now a lot of wells that cost $1 million or more. It has almost gotten to the point where a $1 million well is considered relatively cheap. Years ago, when I first got started with the Humble Company, a $400,000 well had to be approved by the board of directors. Humble was the biggest oil company in the U.S., and still the board had to approve a well that was that expensive.

One of the problems is the dollar is not worth very much anymore. I think a current dollar is probably worth only about 18 percent or 19 percent of what a 1957 dollar was. I know that when I went to Japan in January 1968, a dollar would buy slightly over 360 yen. Just a few days ago, the dollar only bought about 90 yen, and its value is still slipping. That means that the dollar is now worth less than 25 percent of what it was, even in 1968.

That's one of the problems we have here in the U.S., as far as all businesses are concerned: Our Congress has been spending more money than we've got. We have had tremendous deficits every year, whether we have had a Republican or a Democratic president. You can't blame it on the presidents, because Congress controls the deficit. They like to tax and spend and tax and spend until they have literally taxed us into oblivion.

We now have about $6.5 trillion a year in gross national product. Our debt, I think, is currently up to about $4.8 trillion. I can barely think of that many zeros, and if we were actually a business company, we'd be bankrupt. Certainly, as our total debt gets closer to the sum of our gross national product, we are broke. That is why the dollar is sinking in value. That means that as it sinks in value, even in the States, we can't buy with it what we could before. This leads to inflation, and costs go up.

I always seem to come back to politics. We won't have this happen, but I think a benevolent dictator could run the government better than the way our Congress has run it these past forty or fifty years.

The majors will still continue to be the chief suppliers of oil and gas to the United States. Without them, if we ended up with some government agency doing it, we would probably waste half the money that we spend in order to get that oil and gas back into the States. But the majors can bring it in. They have the transportation facilities; they have the refining facilities. They have the various plants that are necessary to clean up the gas, and pipelines to ship it. Or if the gas comes from overseas, LNG tankers would carry liquefied natural gas into the U.S., and thus the major oil companies would continue to do what they have done ever since they got started. They will basically supply to the United States most of its power sources.

The major oil companies have had enough capital to date, so that they have been fairly well self-sufficient. Not to do everything they would like to do, but self-sufficient enough to take on some of the very big projects in the developing countries. They can afford to take the risk in these areas, because they know that if they find something, they will probably find major reserves.

Independents suffer

The little companies and the independents, in general, have not been able to go into some of those same countries, because they don't have enough capital to do that. It's a high-risk deal, and they just don't have access to capital. The banks won't lend them money for a project like that.

That's really one of the big problems for independents these days. Where do you get the money? They may have some reserves in the ground on which they can borrow, but you also need additional money coming into the project, just as you do in any other business, where you go out and promote money from the public. But as the public looks at drilling results these days, it's pretty hard to start an oil and gas company as a new company in the market. There's just no desire in the public to do so. Profits are just too uncertain.

About 80 percent of the independents went broke in the late '80s. I don't think we'll ever see a similar number of independents—in fact, I'm positive in my own mind that we'll never see the number of independents that we had at that particular time. We had approximately 4,500 drilling rigs active in the early '80s, and now, I believe, we have less than 700 rigs active.

The big problem, at a time like that, is that when people suddenly become unemployed, they get into some other profession. The unused drilling equipment then begins to rust, and, finally, it is no longer usable. Thereafter, you don't have experienced people, in case you later want to expand. You don't have the equipment, nor do you have anything else. And you certainly don't have sufficient money in the oil and gas industry, these days, to the degree that it is actually needed, to make it a really healthy, growing industry.

It's going to be tough, but there is always room for a good individual—man or woman. If the individuals are brainy enough, they can probably carve out market niches someplace, where they can still make good money in the oil and gas industry. A lucky few may go on to be much bigger than that and create substantial companies.

As far as the average individual is concerned, there just aren't going to be as many jobs in the oil industry, say for geologists and engineers, as there used to be. However, there are going to be similar jobs in other places. Geologists will become involved in groundwater work. Water is going to be increasingly important in the future, in many parts of the country, where we don't now have adequate supplies, or the waters need to be environmentally cleaned before they can be utilized. Geologists can be helpful in finding where those water resources are located.

Who's in charge?

The individuals that lead the industry will have to be like any other business leaders: they're going to have to have some entrepreneurial spirit; they're going to have to be good financial people. They're going to have to have an excellent knowledge of the technology of whatever their industry might be. Really good business leaders will carry this industry ahead into the future, just as will good leaders in various other industries.

Risk taking depends a lot on the individual. You see people all the time who absolutely don't want to take any risks. If it's a really risky business, they're going to get into something else. They want to keep their status quo.

Successful business people will continue to take risks, but not in the form of wild or foolish deals. In their own mind, they will feel certain that what they are trying to do will be good and will produce results. Those will be the successful business leaders of the world.

You can see that in various industries. There are people in the oil and gas industry who are still building smaller companies into larger ones. Bigger groups are creating new companies many places around the world from ideas conceived about how to make money in foreign countries. That is the type of thinking that must take place, whether a company is large or small.

You may remember what happened in the past to the former Gulf Oil Company. They had so much production and reserves in Kuwait that they thought, "Why should we go look for more oil reserves and take big risks around the world, when we have more reserves in Kuwait than we can utilize in many years?"

Of course, they got nationalized, and they were then in trouble. That's where smart business people would have understood that something could happen to those reserves, that they could be lost. It would be better to have reserves in many additional places. That's basically what most of the oil and gas companies are trying to do now—have scattered reserves around the world so that if they lose some in one area, or if they deplete them, they will still have other reserves elsewhere.

Taking the gamble

A small independent company cannot compete against a major company in an area where it requires a tremendous amount of money. They don't have those big amounts of money. But, if an entrepreneurial company can identify an area that has excellent oil and gas possibilities and can then work out a trade to acquire control of the area from proper country authorities, they can then generally come up with the money necessary to explore and develop that territory.

What total production they develop will probably not be as large as what a major will find, because the majors can go to many countries and have a dozen or two dozen big projects going at one stage, whereas the independent may only be able to handle one big project at a given time. This is what we did in Indonesia. Indonesia developed so rapidly that we had to sell all of our little subsidiaries to come up with the necessary cash required in the earlier phases of development.

With proved reserves, we were then able to borrow money from the banks. We kept getting deeper and deeper into debt, before we were able to get our production really flowing. Then, of course, our project began to pay off rather quickly. But independents generally cannot afford to enter into countries where there are undeveloped and undrilled basins, but no one has any idea of what the basin geology may be like. Independents ordinarily won't be able to put enough money together to step in and do that. That will generally be the major companies' territory.

The greatest risk always is whether you will find any oil and gas. I thought we had an excellent chance to make discoveries in Indonesia because we were in a basin where oil and gas had already been discovered. There had not been any deep drilling, and all indications were that deeper drilling would find more oil and gas. This was the way it happened.

In a lot of countries, one of the big risks that people have is that they might be nationalized. You may get a new government that will not honor the contract the previous one made. They may decide they want to take over the oil and gas properties you have discovered and developed. But in countries where you have only a national or government oil and gas company, they have rarely excelled at what they're doing. The nongovernmental independent companies of the world have been much more successful.

I think that is less of a problem now than it used to be and that more government oil companies will go public or begin to fold up because the countries need their cash and can't put money into operations that are not as successful as they should be. So if they took over, they might produce much of the oil and gas that you found, but pretty soon, they would have depleted it. After that, it is very hard to get money to come back into a country, where companies have been nationalized. You sort of stay away from them. If they change the rules, well, let them stew in their own juice.

In one world

Operations should get a bit easier as we remove trade barriers and restrictions between countries. That's going to happen. GATT [General Agreement on Trade and Tariffs] and NAFTA [North American Free

Trade Agreement] are just the beginning of a wholesome change in trade relations. I used to be a strong advocate of both GATT and NAFTA fifteen to twenty years ago in various international economic fora that I attended. Increasing the ease of international trade is the same type of thing that I was boosting when I was the ambassador to Austria. Let's get rid of a lot of the trade barriers between countries. We really live in one world now. If anybody doubts that, the person should look at what has happened in the computer business in recent years. You can, with faxes and computers, literally communicate instantly to almost any place in the world.

I remember when you frequently had to leave one country and go to another in order to get a line to call the United States. Now, the computer industry, with satellites, software, and literally instantaneous communication, is making a tremendous change in the way international trade can be handled.

I'm not yet properly qualified to step off into cyberspace, but with the various information networks that exist, you can now sit in your office, play with your computer, and pull information out of libraries across the ocean. You can pull information from all sorts of places around the world right into your office.

The amount of information that's accumulating and is accessible in some of the international networks is just staggering to think about. Look what you can now put on one disk. I have recently been at the annual meeting of the American Association of Petroleum Geologists. They have a bulletin that they have published for approximately eighty years. Those bulletins, plus everything else that AAPG has published, I understand, is now on one CD-ROM disk.

So, with the capability of reaching out and pulling in information like that from all over the world, there will be a lot of difference in how businesses will be run in the future.

Lacking the will

I don't think we'll ever have an energy policy for this country. Congress doesn't have the political courage to tackle an energy policy, whether it's about nuclear energy, coal, or just oil and gas. It is absolutely necessary that we pay off our national debt for the good of coming generations. I wouldn't be averse to adding a dollar a gallon to our gasoline tax, or some similar figure, with the stipulation that the money could only be used to pay off our national debt.

Well, most members of Congress would consider that to be political dynamite. I don't think there's a member—or very few members—of Congress who would have the political will to vote to do that, because they would feel that some farmer, some rancher, or even somebody in the city would say, "Gee, that costs me several hundred dollars a year more. I'll throw the fellow out who voted for the tax, because I don't want to pay that much."

Yet at the same time, we pay about 25 percent of the cost for gasoline that the rest of the world pays. In Europe, practically all gasoline costs over $4 a gallon. In Italy, the cost has been perhaps near $5 a gallon. Here in the U.S., in some parts of the country where there are higher state taxes, we may pay up to $1.50, or slightly more. But it's still way under what the rest of the world pays.

We're so darned spoiled, when we compare what we have with what most people in the world have. We sort of expect everything to be handed to us on a platter. The rest of the world doesn't have it that easy. Well, perhaps Europe does, since it is a well-developed area. But if you look at the developing areas of the world, you see that they have to scramble for everything they get.

They basically don't anticipate getting anything out of their governments. Too many of our people have leaned on the government for welfare, for medical care, and for everything else. When you lean on the government for everything, it is a stultifying influence. I used to think a little hunger was a good thing. It made me work a little harder and twice as long as I might have otherwise. I wanted to be sure that I could put food on the table and get married and have kids. If I did, I wanted to be able to educate them and take care of them. Then, once they're educated, well, they're on their own. They can go from there.

But there were many incentives to get out and continue to work. Actually, once you learn something about your business, it's sort of a fun thing trying to come up with new ideas. How can you do your job more efficiently, and how can you render a better, more efficient service for your clients or customers?

You never can tell in this crazy world whether or not you might be dragged into a war someplace. Witness Hitler's rise to power and the tremendous armament machine he built, as he tried to conquer Europe. In a case like that, you need to have energy to power your industry, your ships, your airplanes, and everything else. That's why it's bad to have to import at all. Because if the enemy can cut off your transportation to your country, then pretty soon your plant facilities, your business, and everything else goes down. You can't manufacture. That, incidentally, is another reason that I think we need nuclear power plants. At least you would have electrical power to run some industrial plants, even if you didn't have hydrocarbons.

These problems will all be worked out in due time—maybe with jerks and stops and all that sort of thing. But I have great faith in the conscience and the ability of the people of the world to survive such problems—maybe not as well as they would like, and maybe some won't be solved for a time. But in the end, we will work them all out.

J. HUGH LIEDTKE

retired chairman of Pennzoil Company

J. Hugh Liedtke is a man of the old school. He comes from the tradition in which a handshake is as good as a written contract and he demands the same level of integrity from others. It is this approach that has made him a key figure in the history of the petroleum history.

The former chairman of Pennzoil Company,[1] maker of the country's most popular motor oil, Liedtke has a philosophy degree from Amherst College in

1. Liedtke officially retired on May 19, 1994, after thirty-two years at the helm of Pennzoil. In addition to its motor oil and its Jiffy Lube retail outlets, Pennzoil is a worldwide oil and gas producer and a refiner.

Massachusetts and graduate degrees from Harvard Business School and the University of Texas Law School. He was born in Tulsa, Oklahoma, which was then the center of the oil industry. Liedtke's father was an attorney for Gulf Oil Company, an uncle was an oilman, his father-in-law was a wildcatter, and his wife's three brothers also worked in the industry.

Liedtke began his own career in the industry with his late brother, William C. Liedtke Jr., in 1949 as an independent in Midland. In 1953 he cofounded Zapata Petroleum Corporation.

Through a series of mergers, Liedtke built the present-day Pennzoil into a multibillion-dollar company with diversified, worldwide interests. Despite his enormous success, he is a man not given to extremes and dislikes flattery. My two discussions with Liedtke, who rarely grants interviews, were extremely enjoyable because he has a genial, although reticent nature. And yet this is the man who wrote the book on how to play hardball—just ask the people at Texaco, a few blocks west of world-renowned Pennzoil Place in downtown Houston.[2]

Liedtke made history in the late 1980s. He thought he had wrapped up a deal to buy a large portion of Getty Oil Company, which would have quadrupled the size of Pennzoil. But Texaco, the nation's third-largest oil company, stepped in at the last moment—after the Getty board had endorsed the Pennzoil deal—to buy all of Getty. An enraged Liedtke sued Texaco, and in a decision that stunned corporate America, won an unprecedented $11.1 billion judgment. Pennzoil later received $3 billion in a settlement with Texaco.

*T*he gradual deterioration of any kind of regulation on the price of incoming foreign crude, I feel, has not only been a cause but is the principal cause of the deterioration of the domestic industry.

Various areas of the country have axes to grind. It's a modern version of the Civil War. The Northeast wants cheap fuel oil and cheap natural gas—cheap everything—to run their industry and heat their homes. By having the nation in effect pay the cost of bringing in foreign crude—which to them is cheap individually, but to the nation is quite expen-

---

2. Pennzoil's black, thirty-six-floor, twin-tower structure with slanted roofs connected by a slanted 45-degree glassed pedestrian walkway has been cited by critics as a compelling example of twentieth-century architecture. It was designed by architect Philip Johnson. Reportedly, when Liedtke decided on his headquarters in the 1970s, he demanded that it have a unique shape. The demand for lease space was so heavy that eventually two more floors were added to the original design.

sive—they like that kind of a system better than paying a little bit more individually for domestic crude and perhaps having the country pay less of its gross national product for foreign energy.

And that's true all over the country. Various areas have varying interests in what the price of crude oil and natural gas should be. I started in 1968 trying to point out the need for some kind of a national energy policy. In 1975 I served on a presidential commission to update the Paley Report, which had been written in 1950 and concerns, among other things, a national energy policy. I was a voice crying in the wilderness. Nobody paid any attention. They filed that report in the round file, and nothing was done in all those years.

The United States energy policy is to have no policy.

Short-sighted

What's cheap today is appealing. Nobody thinks about five years from now or ten years. So it's a very difficult thing. I do think the industry should continue to try its very best to educate the public and to educate our political leaders.

We have elected in this country to really not fully develop nuclear energy. We have elected to let the domestic oil and gas industry gradually deteriorate. We have elected to go overseas into a number of different areas—particularly the Middle East, but also other areas such as Russia, and later on it will be Africa. But to a larger extent, these areas require enormous expenditures on our part for military protection. The oil that comes from those areas is pretty iffy, and certainly not cheap, if you look at the total cost for the country.

I know the industry has tried and tried and tried, but I just don't think people want to listen to the message. They do need to get across the fact that foreign oil is not cheap oil, as it is widely described. The cost of keeping the Army, Navy, and Air Force in the Mideast, keeping Israel armed to the teeth, and providing arms to other nations in that area; the cost of rebuilding the Russian oil industry and the Russian economy—all of those things really are at a cost of foreign crude. People should realize that that's an expense the nation has, just as much as the cost of domestic crude.

To some extent, the infrastructure has been destroyed. Let's look at the availability of young geologists and engineers, not just petroleum but civil and chemical engineers. You lessen the demand for them, and then suddenly you need them, and they're not there. You can't train them overnight. That takes several years. The second thing is that we have done virtually nothing in the area of alternate energy sources in this country. At the same time, the single greatest item in our adverse balance of payments is the ever-increasing amount that we have to pay for foreign crude. That's growing all the time and is a drag on our overall economy. It also prevents us, the way things are set up, from developing alternate sources of energy, such as oil from coal sasol projects [a synthetic fuels process used in South Africa] or from shale.

Price stability

As long as the price of domestic crude is tenuous, those projects cannot go forward. You have to have some kind of certainty in the price of crude. If you're building a plant that produces an alternative source because the price of foreign crude is high enough to make that attractive, then all that has to happen is that producers of foreign crude drop the price and the project becomes unfinanceable, and the banks know that.

Banks and financing institutions have been so terribly burned by loaning money on projects in this country, and then having the price of foreign crude go down materially to a point where it becomes uneconomical to produce the domestic crude, that they are not going to finance projects unless they know they can get their money back.

I don't much care how they stabilize the price; some ways are better than others. I'm not an expert in that area. All I know is that it has to be done or there will never be a resurgence of the domestic oil industry, and there will never be, in my view, alternative energy sources developed in this country.

To my mind, we are exposing this country to severe shortages of energy, and it's not going to be possible to play catch-up overnight. We must have energy to run our industry and to preserve the way of life in this country. The development of an energy policy which would assure this country of an adequate supply in the years to come as demand increases should be at the forefront, and to my mind is the principal problem.

We will have to constantly review the cost-benefit relationship between the protection of the environment and the other costs to the country. Some kind of a *reasonable* policy on environmental issues is important. Environmental considerations, some of which are certainly valid and some of which the people—at least of whom I know in the industry—approve of and applaud. But there is some of it that is so far out that it borders on the ridiculous.

That's where some kind of accommodation needs to be made, because there are areas in this country that could be developed quite safely from an ecological standpoint and that would make a material difference in a balance of payments, for example, in the cost of energy. That has to be addressed.

Historically, you get various trends. I do think at this juncture the pendulum has swung too far in one direction. Now it's going to swing perhaps too far in the other, but I'm sure gradually it will come back to the middle. And when the chips are down, and it is a question of man versus the caribou, man is going to opt for man. But I don't think we're anywhere close to that decision at this point.

ANWR needed

I think the failure to develop ANWR [Arctic National Wildlife Refuge] is a crime. I feel certainly it can be developed without destroying the wildlife in the area. I think it would have a very minimal effect. The facts of life are man pollutes because he lives. You're going to have some of it. The question is, is it reasonable? I guess if we had the extreme environmentalists' viewpoint in place, we wouldn't have any oil because the dinosaurs would still be running around.

I know that the industry has taken and continues to take major steps to be in a position to cope with any problem that comes up. But to know the answer to everything before it's happened, I'm not sure they do.

You've got to have some kind of a reasonable balance and I think eventually we'll come to that. It seems to me that certainly we can find one that everyone can accept and that is beneficial to both groups.

My feeling is that there are proper spheres for the government to operate in. One of them is the environment, because that's a national issue.

Minimizing government's role

But I worry a little bit about the government interfering with a good many transactions that take place throughout the world. I wonder how effective they really are in terms of helping. It seems to me they, in some respects, are detrimental.

I think the government participation should be minimal. That much I know. We've had some experiments in the government regulations. For instance, the Federal Energy Regulatory Commission, which, in my view, has been pretty much a catastrophe over the years. We have the government's Department of Energy, which has been pretty ineffective over the years. So I question just how beneficial it is to have the government participating to a large extent in the affairs of the energy industry.

I do think, however, that the government should have a national policy and should be working in that area to try to protect this country's energy needs in the future. Again, I would advocate having some form of energy available to this country which can be produced here in the future.

A policy also needs to take into account the tremendous increase in demand that's going to take place as the underdeveloped nations mature. That's going to occur in the future and we need to have some form of energy here to enable the United States to supply its own energy, not 100 percent, but to supply enough so that it cannot be taken advantage of—either from a security standpoint or from a financial standpoint—at a future date.

One alternative form, of course, is to have the domestic oil and gas industry, even though it's a high-cost producer, continue to produce and continue to be—as a matter of government policy—kept reasonably healthy.

Nuclear power

In addition to that, we need to reexamine nuclear energy. France, for example, has gone a long way in nuclear energy and has standard plants which we evidently do not have. It's kind of like Henry Ford, where you could have any color Ford you wanted as long as it was black. In France, you can have any form of nuclear plant you want as long as it's their single standard. The Japanese are going into that much more.

Perhaps some of the more exotic fuels as sources of energy should be further examined. It's a mistake not to be looking for some alternative to a foreign supply. It *can* be cut off, even today, in a world that seems to be troubled everywhere.

If Nigeria is cut off, the price fluctuates madly. If the Middle East is threatened to be in trouble again, if there's a big strike somewhere, those affect things so markedly. It's unfortunate. But if we allow ourselves to have no alternate source of energy, and the domestic oil and gas industry is pretty much put out of business, and we have not developed an alternative to it such as nuclear or some other form, then someday, those who have the power to control the price are going to really put it to us, I fear.

Easy target

I don't know what the oil and gas companies can do to help in a situation like this. I almost despair in that I think they've tried awfully hard to educate the public, but the industry is such an easy target for the media to attack and, generally speaking, it seems the media much prefer something sensational as opposed to educational.

The major companies will continue to be in the forefront of developing oil and gas throughout the world. There are about a half dozen of the really big companies, and they have the capital structure and personnel to carry out that mission. Unless something is done to stabilize the price and some reasonable accommodation is made on environmental issues, I think they will continue to go overseas more and more.

I feel there needs to be more government cooperation with American companies abroad. Certain other nations—the British, for example—seem to be politically in touch with companies like BP [British Petroleum] or Shell throughout the world constantly and operate together with them. That's not so true in the case of our country.

It would be very helpful to the nation if it [government] were more cooperative. The problem is, at least from the statistics I see, that just past the turn of the century, demand growth will probably be such that it's going to be very difficult to fill with just oil and gas.

Whether the majors return to this country I really can't say, because quite frankly, the foreign areas are the least explored and they need huge amounts of oil. At some point they probably will. It may be fairly far out, after the big obvious structures and basins and so on are developed abroad, and oil gets really tight, then I think they might come back here and start going after oil that they know is down there, but currently cannot be produced or, if produced, is very high cost.

There are hundreds of millions of barrels trapped underground, for instance, in West Virginia. Probably less than 20 percent of the oil in place has been produced. The gas was blown off into space before the turn of the century because nobody really understood the purpose of gas—what it did, the energy source that got the oil out of there—so a lot of that oil is down there. But it's expensive to get out at this point.

With all of the majors moving overseas and the current fashion of core areas, they are selling off a lot of properties in various areas. This is perhaps the biggest opportunity that domestic producers have had in some time in the sense that if they can finance their projects—which is the big thing with an unstable price—the majors are farming out and dropping things that independents couldn't touch a few years ago.

So they do have a chance to get hold of properties that they can probably operate cheaper than a lot of the majors. One person can make a decent living out of one or two good oil wells. The problem is in financing sizable projects, tertiary for example. It's hard to finance a big tertiary project as an independent, because a bank doesn't know what the price is going to be two years from now, so they're very reluctant to loan.

Downsizing

Unfortunately, people come into corporate lives to believe that a corporation is really going to take care of them, kind of the paternalistic way that they used to operate. But that's really not true and I think it comes as a shock to a lot of people who are totally unprepared for it.

We don't have the Japanese theory that you keep people on regardless of whether things are going well or badly. That's not just true of the oil industry, it's all of our industries. But it comes as a shock to people when they have to be zeroed in on, and that's going on all over the country today in virtually all industries.

I would hope, and it's not happened, that the younger generation of oilman, particularly independents, would find some very articulate spokesman. I think the employees of the major companies as well as the independents have a personal financial interest in trying to educate people and to have a spokesman who represents them.

I am not sure that some of the large industrial groups are very effective in Washington. In fact, the endorsement of some things by certain ones of them seems to me to be the kiss of death politically. So I would hope that some kind of leadership could come from some of the younger people, who are particularly in the domestic industry.

I do think there will always be a place for the entrepreneur. There's always room for an entrepreneur in this country. It becomes more difficult, but it's always possible. For one thing, it's difficult to go very far because most of the substantial companies have a long history to them. People don't seem to realize how much time means in terms of trying to build something.

[The hallmark of the industry has always been a man's word and a handshake. Liedtke was asked whether the Texaco case is symptomatic of what has happened to the industry.]

No, I really don't think that. It is true that a man's word is his bond. I belong to the All American Wildcatters—I don't know if you've ever seen any of their stationery—but that's right on the legend of it, the seal they have. But no, Texaco was different. I don't think most companies were like that. I really don't. I think most companies will do what they say they're going to do. Texaco, at that time, was the most unpopular oil company in the oil field. You can check that with any number of people. Just ask them. But it's beating its way back now. I think it's better than it was. Maybe they learned an expensive lesson, I don't know. It was damned expensive.

I felt the thing was pretty overblown. It was a simple case in contract, really. A pretty obvious case. Hell, a judge in Delaware said it when he wouldn't grant specific performance, that we had a contract. I think the thing that was hard for people to understand was the amount of damages, and I don't think you'll ever make them understand that. But the facts of life are that it had taken a hundred years to build Getty Oil. You don't pay for that kind of thing. It's invaluable. There's no way that one man in a lifetime can build another Getty Oil.

So it's a funny thing. Everybody thinks that $3 billion settlement was a magnificent victory. I've always regarded it as a defeat, because what we were really after was Getty Oil and not $3 billion. Getty Oil would have made a real difference. Three billion dollars makes a difference, but I don't know where you can get another Getty for $3 billion. If you're trying to build an oil company, it's not the answer.

Majors

PHILIP J. CARROLL

president and CEO of Shell Oil Company

Philip J. Carroll is president and chief executive officer of Shell Oil Company, the nation's leading gasoline retailer and the largest operating company of the Royal Dutch/Shell Group, one of the world's greatest energy companies.

Carroll joined Shell in 1961 as a petroleum engineer and worked for the company in his hometown of New Orleans. He then worked in Texas, New York, and Los Angeles before going to Washington, D.C. in 1973 as director of the Energy Conservation Division in the U.S. Department of Commerce and as executive director of the National Industrial Energy Conservation Council.

In 1974 Carroll returned to Shell Oil, and in 1978 he was named general manager of production, Western E&P Operations. He was elected vice president of public affairs in 1979, and in 1985 he was assigned to Shell International Petroleum in London as managing director of Shell International Gas.

Carroll came back to Shell Oil in Houston in 1986 and was elected to his present position in 1993. He holds a B.S. degree in physics from Loyola University (New Orleans) and an M.S. degree in physics from Tulane University.

During a discussion at Shell Oil headquarters in March 1995, Carroll, a warm and gracious host who takes a genuine pleasure in helping his visitors feel at home, provided an absorbing account of upstream and downstream activities on the domestic front.

A number of tough issues face the oil industry as we finish out this century. Economic, political, environmental, legislative, regulatory, and technological concerns continually complicate our efforts in the energy business. I'd like to discuss these issues by focusing on the upstream and downstream sectors independently, as the challenges in each arena are very different.

First, let's discuss the upstream business, which includes exploring for, finding, developing, and producing oil and natural gas resources. One of the biggest challenges facing domestic upstream activity today is the maturity of the province. More wells have been drilled in the United States looking for and producing oil and natural gas than in the rest of the world combined. Our geologic basins have been sampled and tested, and the prospects of making significant discoveries in the older, more mature areas lessen every year. Yesterday's small discoveries are today's big finds.

The United States is now the world's highest-cost producer. The cost structure associated with finding, developing, producing, and selling a barrel of oil in the United States will likely always be higher than in the Arabian Gulf or in lightly explored areas such as Mexico and Latin America. This is attributable to such factors as province maturity, location, and quality of the hydrocarbon accumulations, combined with the cost of labor and amount of regulatory overhead associated with drilling and producing.

Add to that any efforts of the world's low-cost producers to increase their market share by reducing the price of oil—as we've witnessed many times since 1986—and we understandably see a dramatic decline in U.S. exploration for oil and gas, both in numbers of companies participating and in levels of investment.

Another significant challenge in the upstream is gaining access to explore potentially hydrocarbon-rich areas. The vast majority of the lightly explored areas are in the offshore Outer Continental Shelf or in Alaska. Exploring and developing in these areas can be expensive due to the location or terrain, but the greatest challenge is that the preponderance of prospective areas are off-limits due to environmental concerns.

We have essentially the entire offshore areas of the West Coast and East Coast of the United States off-limits, as well as the eastern Gulf of Mexico adjacent to Florida. This ban on exploration and development also extends to certain areas of offshore Alaska and, of course, the Arctic National Wildlife Refuge.

Even those of us who bought leases in many of these areas are not allowed to explore or develop them. This has been a frustrating situation indeed. To correct this problem, we will need all parties, including industry, environmental groups, and legislative and regulatory bodies, to work together to find credible solutions for non-destructive, cost-effective exploration and production in environmentally sensitive areas.

One bright spot in the domestic E&P [exploration and production] picture is the central and western Gulf of Mexico, off the coasts of Louisiana and Texas. This area has a long history of exploration and development, and the federal government representatives of this area as well as the state governments of Louisiana and Texas recognize the economic benefit to their citizens. But, of course, economic benefits are not the full picture: the state governments also recognize the excellent safety and environmental records of the industry. It's a win-win situation for all involved.

The Gulf of Mexico appears to hold promising large prospects, particularly in deeper water. We've begun producing some deep-water fields, and have several other prospects under development. The expected contribution from deepwater prospects to our nation's oil and gas supply will be substantial into the next century.[1]

However, deepwater efforts alone can't reverse the steady overall decline of production and investment in E&P in the United States. This has several negative ramifications: increasing dependence on foreign oil (and associated national security and environmental issues), balance of payments difficulties, employment impacts, and a lessening of America's technological leadership and capability as we do less domestically.

While all of these are painful consequences, one of the more troubling is passively allowing yet another of the technology areas in which America excels—an area which has translated into advances in other fields, such as the medical and computer industries—move overseas to be refined and advanced while America lets the knowledge and its associated benefits die on the vine.

Our country can't afford to let this happen. Instead, we need to encourage exploration and production domestically by opening up areas currently off-limits, reducing unnecessary regulatory burdens, and establishing policies which stimulate development of marginal properties if deemed to be in the nation's best interest. The nation will benefit from the economic stimulus as well as from technological improvements.

Refining and marketing

Turning our attention to the downstream, I feel the United States is a wonderful market for manufacturing and selling petroleum products. We are the automobile society (although there are some parts of the world anxiously trying to catch up with us, especially in developing countries). But the refining and marketing industry in the U.S. has been

1. Production from the Auger platform began in April 1994 and represented a landmark in offshore development. Constructed at a cost of $1.2 billion, the Auger signifies development in the deepwater Gulf, which might be the last great frontier for the United States. Auger is located 135 miles offshore in 2,860 feet of water. It is a water-depth record for oil and gas production in the U.S. and is the largest tension leg platform ever built. A TLP floats on the surface and is anchored to the sea floor by tendons. Auger has produced as much as 55,000 b/d and set a Gulf record of more than 13,000 b/d from a single well.

particularly troubled over the last several years. In truth, if we look back over the last ten to fifteen years, we'd find that it has been a business where few of the competitors have been able to earn their cost of capital—and that's an unhealthy situation.

Increasing regulatory requirements have played a key role in the industry's declining return on investment. For many years, society has been calling for cleaner air and water, and these requirements have been translated into various restrictions and regulations which have helped to improve the quality of life for all of us. The emphasis has shifted in the last five years or so, however, from what the required outcomes should be (for example, reduce emissions to x percent) to specifically how to achieve those results, including what components should be included in the gasoline blend.

There are numerous problems with this approach. Using the Clean Air Act and the ethanol mandate as a case in point, the overriding problem was that the most sound, scientifically based solution was not selected. Related to this was the interwoven political agenda which serviced the desires of corn growers and ethanol producers to gain market share, even after studies failed to show that ethanol was any better for the environment than what it replaced.

Rounding out the issues list was the need for ethanol to be subsidized to make it cost-competitive. When the facts were reviewed, it was apparent that the driving force was more political than environmental or economical. The mandate was aggressively contested by the industry and others, then was blocked by the courts.

On a broader scale, requirements related to the Clean Air Act brought refiners and the public a lot of frustration. As new requirements were enacted and some of the associated expenses were passed along to consumers, the public began to question their cost/benefit. State and local governments, feeling the heat, began to push back on environmental regulations. Many areas that were in programs voluntarily began to opt out. Therein rose another problem: the refiners had already spent billions of dollars to enable them to manufacture what the government said they must, yet the demand for that mandated product began to fall away.

More appropriate ways to address these environmental issues would be to delegate more responsibility for regulation to states and municipalities so that legislation doesn't try to be a one-size-fits-all answer; to let the regulators set objective, results-oriented requirements while enabling the scientists in the industry to determine the best ways to meet those requirements; and to allow the free market mechanism to work without special treatment for pet fuels or components. I'm encouraged by the healthy impact that push-back by the public and local/state governments seems to be having, including energized discussion about scientific validity and cost/benefit proof for HS&E regulations.

Tort reform

Another area of high cost to the industry in both the upstream and downstream is the unrestrained use of the civil justice system. It can stifle innovation by causing companies to refuse to bring to market useful new products because they feel that potential tort problems would be ruinous. It can cause good companies to leave good businesses altogether.

No one wants to take away an individual or group's rights to just compensation for injuries or harms that have been done to them. Certainly no one wants to take away the power of the court system to punish people or companies who act in irresponsible ways. But the truth of the matter is that many companies offering and marketing safe products and acting in the most conscientious and scrupulous ways still find themselves subject to incredible judgments based not on actual circumstances, but on ability to pay.

We've seen in the last twenty to thirty years a degradation of our criminal and our civil justice systems, which has made our country a laughingstock to many around the world. The tort situation is a very important factor and reform is an absolute necessity if we are to stop a rampant problem. It is critical to making this a better business climate.

Restructuring

The challenges in the upstream and downstream sectors of the oil industry led to the extensive restructuring we've seen over the past several years. Companies have been and are continuing to dispose of assets, acquire assets, downsize, and reposition themselves in ways which significantly alter where and how they operate to make a profit.

In the upstream, properties are sold to competitors who feel they have a competitive advantage. In the downstream it includes shutdowns or partial dismantling of refineries—anything less than a very efficient, well-equipped refinery is too much of a cash hemorrhage. It also involves withdrawing from certain markets where a company has been selling products, then reentering under a different basis or in a different way. We've done this ourselves, when we sold all of our service stations in the Atlanta market, and later came back by association with some very successful jobbers that carry our brand. This whole process of restructuring is one I believe will continue for many years.

Given all the issues which face the oil industry, how can a company expect to compete successfully? While we will continue to influence external matters in ways open to us, at Shell we realize that we must focus on the many things we can change ourselves—transforming our company and our people to position us to succeed regardless of the curves thrown at us. Like others, we are exposed to fierce global competition, and information technology has made it possible to work in wholly different ways. The same thinking that worked when we were riding the heady waves of the '70s and early '80s won't take us where we need to be anymore. Success requires transformational change. In some cases, it is necessary to lead the organization to creatively destroy what has led to past success. Cost reductions and downsizing have served to initiate this process. These were enormously painful for all of us, but the blunt fact is that we had no choice.

The time has come for rebuilding the organization, but in a much different way than it previously stood. A new relationship between company and employee is being built based on mutual commitment rather than loyalty and security. For the employer, the commitment is to provide the freedom and opportunity to contribute; to reward results, both financially and psychologically; and to provide the means for employees to learn and develop marketable skills. For the employee, the commitment is to apply their talents and energy to the success of the enterprise and to take responsibility to increase their contribution over time. This mutual commitment leads to security owing both to the success of the venture and to the employee's confidence in their value in the market-place.

But learning goes beyond individual learning. I believe the ongoing success of corporate transformation depends on organizational learning. Like individuals, companies have a collective sense of identity, a sense of fundamental purpose, and a capability to learn from the past and present. Both individuals and organizations must be capable of continual learning to ensure the success of the company.

How will it feel to work for such a company? Exhilarating and scary. Change is high-octane fuel for our intellect and our emotions. At the same time, it is an imperative if we as a company intend to be around to achieve success in the future.

VICTOR G. BEGHINI

president and CEO of Marathon Oil Company

It seemed natural that Victor G. Beghini would go to work in the petroleum industry. Growing up in Greensboro, Pennsylvania, the future president of Marathon Oil Company often helped out at his father's service station—changing oil, pumping gasoline, and becoming well-versed in the concept of customer service.

Beghini also had an aptitude for engineering, which together with his love of nature led him to the Earth and Mineral Sciences School at Pennsylvania State University. He received his B.S. degree in petroleum engineering from Penn State in 1956 and soon went to work for Marathon.

Marathon was the former Ohio Oil Company, established in 1887. It was the largest of Standard Oil's producing companies before being dissolved by the Supreme Court in 1911. Marathon fought off a takeover bid by Mobil Oil in the early 1980s by agreeing to be bought by U.S. Steel, now USX Corporation, for nearly $6 billion.

Beghini worked his way up through the ranks of Marathon and in 1987 was elected president of the integrated company. In 1990 he was elected vice chairman of energy and a member of the board of directors of USX Corporation. He is a member of the National Petroleum Council and the All-American Wildcatters.

Today, Beghini is one of the most respected leaders in the petroleum industry. In interviews, he is typically outgoing, good-humored, and forthright in discussing the vital issues and challenges that confront his industry.

There's a broad spectrum of issues facing our industry today, but I think the three areas of greatest exposure relate to technology, capital, and politics.

Technology must continue to drive this industry into the future. As the easier discoveries are made, future discoveries that will move this industry forward are probably going to be in more hostile, more remote environments. Therefore, technology will be more critical if we are to discover new reserves and bring them to market.

By hostile environments, I mean deep water or where you have severe weather conditions. The more remote the prospect, the bigger the accumulation of hydrocarbons you must find to justify bringing it to market. And it goes without saying that you must operate in these difficult areas without any detrimental effect on the environment. So, the footprint of the operation must be considered along with its difficulty, and these objectives can only be achieved through greater technology.

From the standpoint of finding new reserves, I don't have any doubt that there will be further enhancements in seismic. Three-D is the boon now. Some people are talking about 4-D, which basically is 3-D run in a sequential time basis to where you can run 3-D seismic at varying times in the field life, and you can maybe see fluid movement or see depletion or see what areas of the field are participating in depletion.

I also think you'll see technology advance in terms of where we can produce. It wasn't too long ago that people thought 500 feet of water was deep. But now people routinely drill wells in 2,500 to 3,000 feet of water without thinking too much about the technology required to do that.

In my opinion, you're also going to see production technology evolve. There's a lot to learn in terms of the displacement process in the reservoir, and I believe that further understanding of that process will lead to enhanced recovery in older fields. For example, three years ago at our Yates field in West Texas, our production was down to a little over 47,000 barrels a day. With the advent of a new displacement process utilizing short-radius horizontal drilling, that field has now seen rates as high as 55,000 barrels a day. Yates was discovered in 1926, so I'll call it a middle-aged oil field because it's probably got another fifty or sixty years of life, thanks to creative engineers and their application of new technology.

In the downstream side, there's just as much opportunity for technological advancement. I think you're going to see new ways to process heavier and sour crudes that will be more efficient. I think you'll see the ability to process cleaner and better products come about. So there's a lot of ways new technology will serve our industry. The fact is, this business can't exist without new technology.

Increased efficiency

If you look back at all the downsizing the industry had to undertake in recent years, I believe you will find that downsizing is also a companion to technology. At today's prices, downsizing had to take place. But, as computers have become more and more proficient and as software has become better, geologists nowadays can look at the 3-D seismic they put in a workstation, play with it, and draw maps digitally. So one geologist may be as productive today as six to eight geologists were fifteen years ago.

The same applies to engineers. Our pumpers today in West Texas walk around with hand-held computers with which they can run a test on a well. They punch it into the computer and the information can be in the hands of an engineer to analyze that well test almost in real time.

So, you don't need a lot of people to write and prepare reports. As you handle data once, you don't need the bodies to massage and type and re-form and reformat data.

I believe that downsizing was inevitable. Price was the catalyst that caused it to occur. But I think good companies would have downsized anyway, because technology allows you to do so without sacrificing productivity.

Another key question is, What will it take for people to continue to invest in the industry? The answer is simple: financial return. The opportunity for T-bills is always out there. So if you can't get financial return, there's no motivation to invest in this industry. That's also true for individual companies. I believe that where you can get the return, those are the countries you'll go to. Where you can't get return, those are the countries you will leave. Decisions will be based entirely upon that, because the industry will have a crying need for capital.

We have to face the fact that with every passing day the known reserves in the world become more concentrated in one area of the world, that being the Middle East. So, as we move forward in time, we have to ask ourselves, What are the political ramifications long term in the Middle East?

Within those ramifications, we must also ask, What capital requirements are Middle East governments going to put forth to bring new reserves on? And, Does this provide an opportunity for private enterprise? I believe it does, because we're now seeing countries like Iran talking to private enterprise after avoiding it for quite a while. I don't have any doubts that Iraq would welcome private enterprise today, both for capital and technology. In spite of this concentration of reserves, however, capital will still be required to develop them. And I think this is an uncertainty right now.

Where will this capital come from? The lenders of this capital will have to be convinced that they are loaning this money to someone who not only has an opportunity to pay back the loan, but can also make the risk equivalent to some non-risk-bearing kinds of investments. My point is that you're going to have to have a set of economics that draws capital into the industry. Right now that set of economics is not there. That's the second very broad issue we're facing.

The third issue is the big unknown—political exposure. You only have to look at the politics in this country to recognize the degree of the exposure and the variances of politics. We've gone to a position in this country where the only thing open for drilling is in the Gulf Coast off-shore. Essentially, everything else in the country has been shut down. We are a country that imports over 50 percent of our liquid fuel requirements, yet we don't seem to be able to come to grips with the fact that our domestic reserves can be developed in an environmentally safe manner.

We need a political change in Washington, where the mindset is that we cannot explore. We need to change that to a policy which puts forth the basis for exploration, recognizing the environmental risks, and then lets companies who want to accept those risks and sensitivities have an opportunity to explore. Even with all the romance associated with the many areas industry wants to explore, we still don't know whether drilling would be successful. The Gulf of Alaska was touted as the next big North Slope many years ago. I think there were fourteen wells drilled up there—all dry. Just because we think something is there doesn't eliminate the risk. There might not be anything there, or the fields might be so small they can't be brought to market.

Compounding our problems is the fact that there is no economic basis for our government's regulatory posture at this point in time. Many regulations have no economic or scientific basis. They were drafted and put forth in an era of extreme emotionalism. Someone is going to have to go back and bring rationality into the regulatory system. It's not that it can't be done, because there are governments around the world that have done a very, very good job of working with our industry.

The one that automatically comes to mind is the United Kingdom. The United Kingdom has basically changed its tax laws in order to maintain a very active, viable energy industry. They have used their tax laws as a vehicle to promote economic growth and employment in the oil sector. When prices went south, they changed their taxes to make it more feasible for companies to develop smaller fields. They did this because there is a latent employment effect over there that is very beneficial to the country in terms of tax revenues.

What they give up in direct energy revenues, they make up from the workers who would be unemployed if it weren't for the energy business. And the U.K. has fostered this pro-oil philosophy without compromising environmental and safety considerations.

We're also seeing countries all over South America privatizing because their leaders recognize that our capital and our technology are great assets to tap for their own benefit. They're smart enough to realize that a healthy energy business is good for their countries. And again, they can benefit from energy development without any degradation of the environment.

This country, on the other hand, is moving toward what I call an "off-oil" agenda. That agenda by itself is not economically feasible. In fact, it's economically deficient, particularly when you consider that our government is willing to subsidize fuels in order to move this country off oil. The economic choice of fuels is a very, very important choice that a user should have, whether that be a factory or an automobile owner or anyone else who uses fuel, whether it be natural gas, methanol, crude oil, coal, whatever. For government to mandate whether one fuel is favored over another does nothing but compromise the economic health of the country.

Can the political landscape in the U.S. be changed for the better, as in other countries? Yes, I believe it can if we remove a lot of the emotionalism from our discussions and incorporate good science and good economics. Then, we'll find out that there's a lot of areas of accommodation between industry, environmentalists, and consumers. I'm not too sure there will ever be an accommodation among the more radical sides, but I don't believe some of those individuals and groups should have a place in an environment of rational discussion anyway.

Cost-benefit analysis

One way to realize needed political change would be for Congress to pass a law, such as the one almost passed in the last session of Congress—it was passed by the Senate ninety-five to three. This law should incorporate into regulations a cost-benefit analysis before new regulations are issued. That would be a very, very helpful thing for all U.S. industry, not just the energy business. And such analysis should not apply only to new regulations. Just as tax policy was made retroactive last year, all existing regulations should have to be reexamined and justified in terms of a cost-benefit analysis. You can't have economic growth and you can't be competitive globally if you don't do this.

Cost-benefit analyses should also be applied to any regulations requiring alternate fuels. If you're going to put economics into the equation, there's very, very little out there that will compete with oil and natural gas. There's a lot of talk about electric cars driving up and down and around Houston and not creating smog. That's a beautiful thought. But people ought to think about where the cars are going to get their power from. If you're going to have a power plant that burns coal or something else to get the electricity, that will create emissions as well. And, on a mile-driven basis, those emissions are probably going to be more than they are with today's automobile.

Today's problem in this country is not the car and it's not the fuel. Today's problem could be easily solved with two things. And those are a rigorous inspection and maintenance program and some recognition that old cars should be taken off the road. Let Marathon or any other company that has a mind to acquire those old cars. Detroit certainly should not be viewed as a villain, either. Cars now are much cleaner than they were in the '70s. In fact, we are probably taking out over 95 percent of the emissions compared to what we were doing in the late '60s. So, it's not Detroit and it's not the fuel.

What we have is a government that practices command and control. What you're looking at in Houston right here in this community is ride-sharing. Ride-sharing may be what this community needs, because it would finally show every person in this community just how much government can really intrude on his or her choices of what they want to do.

A California newspaper, for example, reported that if you use our government's own figures, employers are looking at an annual cost of $232 per employee per year to implement an effective ride-reduction program. This means that a company with one hundred employees will spend $70,000 over three years to remove 154 pounds of emissions from our atmosphere. At the same time, it is estimated that 10 percent of the older, poorly maintained cars on the road today account for upward of 50 percent of all auto-related pollution. This means that an employer—if given the option—could buy two clunkers for $1,600 and achieve the same pollution reduction that will cost a currently projected $70,000.

So I have to ask, What is the ulterior motive? It's obviously not to clean the air. If you wanted to clean the air you would say, let's buy up some old cars. Or you'd give every company of over one hundred people in Houston an emission level. And you'd say, get down to this emission level. We don't really care how you get down to it, just get down to it. That would have a much greater impact. It would be much more sanitary, and people would still have the freedom to choose how they get to work. More cost-benefit analysis, less command and control. That's what our industry and country need.

Access

A second policy I advocate would be an opening of public lands to multiple use. This should be done with environmental regulations in place that are appropriate for the type of activity. Most companies in my industry are willing to live with those kinds of regulations. If companies assess the opportunity for discovery is greater than the risk, they'll do it. If they sense it's less, they won't do it, no matter what happens. But there's no basis other than fear to automatically just close areas such as the entire coast of California, ANWR in Alaska, and large parts of the East Coast. That's silly, because when you look at spills from offshore platforms, they're minuscule compared to spills from oil tankers. But regulators are saying we favor a policy of being supplied by foreign tankers rather than being supplied by offshore oil platforms or other domestic onshore activities. From an environmental view, that makes no sense at all.

The first area that should probably be opened to exploration is ANWR. It has a lot of potential. There's no real logical reason to shut it off when you look at the size of the footprint of the operation. In other words, the number of acres that might be looked at for an operating purpose is ridiculously small compared to ANWR's total amount of acreage.

There's been a lot of talk about the effect on the caribou herds. All you have to do is look at the North Slope. You will find out that the animals have multiplied significantly in that operation's environment. And any operation in ANWR would be better from an environmental standpoint simply because it would be newer. Technology has allowed us to leave a smaller footprint as we go out in time.

There's a very real problem associated with ANWR because, at some point, with depletion of the existing fields up there, the pipeline from the North Slope to Valdez is going to have increasingly unused capacity. And at some point, that pipeline is going to shut down because the tariffs get too high. It would be rather ridiculous to have that happen, and then have to go back up there to restart and rebuild large sections of that pipeline. That makes no economic sense. So the ANWR is the first place that should be made accessible for exploration.

The second place might be off the East Coast. Marathon has some acreage off North Carolina that the federal government awarded to us in a 1981 lease sale. But it hasn't allowed us to drill on it since then. That in itself is not so bad, but the federal government has also seen fit not to give our money back. We believe that's taking without due process. In any other kind of environment that would result in a lawsuit being filed. Not only would that entity probably be found guilty for actual damages, but there would be punitive damages found, because the federal government's actions border on fraud.

I believe the East Coast is an area of opportunity. There are also promising areas in the eastern part of the Gulf of Mexico and offshore California, although the entire state waters off California have now been closed to development. There are other areas in Alaska and in the Rocky Mountains.

If this country were more open to drilling, there's no doubt that everybody would be drilling more wells here today. Economics being what they are, I don't know that the count would be significantly improved. But, if a lot of these areas had been opened to drilling ten years ago, you'd know by now whether there was anything in ANWR or not. You would have drilled off the coastline of California—which, by the way, is the nation's largest statewide user of energy. There are a lot of lands throughout the West that would have been bid on. Marathon would be active bidders in those areas.

We would have had our well drilled off North Carolina, and maybe we would have found something major. And perhaps it would have been natural gas, which really is interesting because what environmental pollution is there from natural gas? So, I think you would have seen a much higher level of domestic activity in the industry in general, had industry not been excluded from those areas with the greatest promise.

I want to reemphasize that I am not saying "to hell with the environment." What industry is saying to government is "you draft the rules." You make the consequences of environmental degradation such that we will decide. And if you make those rules so punitive that we believe the risk of hurting the environment is much greater than the risk of economic discovery, you probably won't get anybody to come talk to you. But there has to be some rationality in the rules.

In California, for example, Exxon has fought for over ten years to get the Santa Ynez field on line. They've finally done it with offshore loading only because the State of California does not want them to bring the oil ashore by pipeline. Now that's really silly. To offload in a tanker rather than bring it to shore in a pipeline makes absolutely no environmental sense at all. As I said, it goes back to emotionalism. It's not rational.

Obviously, there are environmentally sensitive areas. But when you consider some of the areas in southern Louisiana, for example, you find the Audubon Society owns lands that have producing wells. I believe those areas too sensitive for our operations are going to be few and far between. For example, Marathon drilled in a very sensitive area in the Rocky Mountains near Yellowstone Park about nine years ago. We had to helicopter the rig in. We helicoptered all supplies in. The well was dry. When we got through, we planted buffalo grass. I doubt that there was any environmental degradation at all as a result of our drilling that well. The fact is, there may have been an environmental enhancement, considering some of the scrub that was up there. The U.S. Forestry Service and Bureau of Land Management even issued Marathon a commendation citing our exceptional efforts.

A number of other projects we've completed over the years have also been recognized by federal and state agencies as being beneficial to the environment. We have done things like building Prairie Chicken habitats in New Mexico and putting screens over our stacks to ensure birds are not harmed. Emergency preparedness drills have been conducted at almost every major operation, and local emergency response organizations and officials have been involved in these actual drills to ensure the best response possible and to minimize the effect of a possible incident. These kinds of things are being done by all companies, not just us.

I don't really know whether a company can set too high a priority on environmental issues today. They have to be high. That's why we have environmental principles that we live by. These principles basically state that we're to conduct our business within the laws and regulations of the entity that we are located in. That we respect the environment. And that we will do whatever is necessary to make sure that we do not degrade the environment with our operations.

We have an environmental engineer get involved in almost every major project we do. We build safety and environmental considerations into the original design of our operations and activities. Then we have an environmental and safety organization that audits our operations. In fact, in 1991 we were recognized as the safest company in the industry from the standpoint of lost-time accidents. In 1992 we had an off year, but we're back now among the leaders in the industry. If you have a workforce that thinks about how to do things safely, they are basically thinking about how to do things that are environmentally safe as well. If they think about how to do a job without hurting themselves, they're not going to hurt the environment either.

So I firmly believe that there are ways to do things in our business that will not harm the environment. Nobody wants to harm the environment. We all live here. But there's a tendency to go overboard from the regulatory side. I would prefer for the government to set regulations and give us the opportunity to decide how to meet those regulations. In other words, get away from command and control and let creativity of the individual and technology solve the problem.

A study done at an Amoco refinery in Yorktown, Virginia, provides an excellent example of what I mean. Amoco offered a plan that would have multiple times the positive environmental effect over what the government regulations would accomplish, for one-fourth the cost. And the government agreed that Amoco's plan would have multiple positive effects. Yet the government told Amoco, "You can do those things if you want to, but you still have to do what we're mandating, even though what we're mandating will not have near the positive impact of your own program." Now those are the kind of regulations that border on the Neanderthal to me. I can't imagine any government coming to that conclusion after a group of experts gets together and agrees that another solution would be much more beneficial.

Tax policy

A third area needing change is our tax policy, which is less than generous to industry in general. For instance, the alternative minimum tax is a ridiculous tax to place on industries that are going through bad times. It's silly to have to pay taxes to the government when you are having an operating loss. That may be changed. There's an opportunity to do that.

Specific to our industry, I've been very active in trying to put a safety net under marginal wells. Ninety percent of the wells in this country are marginal. We're not asking for a tax benefit that runs on. What we are suggesting is the development of a safety net, so that when prices begin to drop to a level where some of those wells may have to be abandoned, there is an opportunity for a tax credit to come in and keep those wells operating.

The State of Texas and the Railroad Commission and the Texas Legislature have done an outstanding job in looking at incentives for producing wells, both oil and gas. And they will tell you in fact that such incentives have had a very positive effect on revenues. They not only keep the wells producing, but provide employment. Many of the wells in old fields probably will only recover 20 percent to 30 percent of the oil in place. If we plug those wells, the rest of that potential recovery is probably gone forever. If we keep those wells operating, then maybe as new technology comes on, it will have a utility in that particular field that will allow more oil to be recovered.

Again, such incentives provide jobs, increase our producing rate, and provide taxes to state, local, and federal entities. We're not asking for a big tax break. What we're saying is, give us a safety net when times get bad. When times are good, we really won't need the safety net.

This would be a positive income factor to both government and local economies.

As for any oil import fee, you will find most people split, not because it's an import fee per se, but because politically it cannot be passed. The first thing the Northeast coalition of senators and representatives would want to do—and forget the political party—is to exclude fuel oil, because we import large volumes of fuel oil and their constituents heat with fuel oil, thereby putting the Northeast at a disadvantage. They say, "Okay, let's exclude fuel oil." And somebody else says, "Well, how about gasoline? Our constituents have to drive their autos a lot more miles than in other areas of the country."

Nobody can guarantee that if you get an oil import fee through Congress, that it will be an import fee on all imports. There will be exceptions. Those exceptions really make the oil import fee a Trojan horse. There will be benefits that will not be compensated for. Many people want an import fee to drive up the cost of domestic production. Well, my own feeling is, why do you want to drive up the cost of domestic production or the cost to use domestic production? That really does not help our world competitive position, particularly for energy-intensive industries.

So, I think you would find that even with that negative, an oil import fee probably would be supported by a wider span of constituencies, if in fact it would apply across the board for all products that came into these United States. I don't think that has any chance at all of passing. I don't know that we really need that kind of help. I believe we would be better off having the government provide a safety net. When times are good or when economics are viable, let us make the decision whether to drill or not in this country. We don't need government help to do that.

Furthermore, I don't think our industry really wants or needs the kind of attention we'd reap as the result of an import fee, which would raise everyone's cost of energy. Politically speaking, one of our problems has been that we've probably gotten more attention as an industry than any industry in this country over the last eight or nine years. And when you look at it, very little of it has ever been positive. It's all been negative. It's all been on how we want to bring this industry to heel. It's how we're moving this industry into a smaller environment. And it has had its effect. When you lose 500,000 jobs in twelve years, that's a significant employment effect. That's more than the combined automobile and steel industries lost by a factor of two, I believe.

So, we've gotten a lot of attention, and all of it has been bad. What we would like to get—compared to other U.S. industry—is the same attention, respect, and recognition that energy is very basic to the economy of this country. If you accept that premise, then you have to ask yourself, How can this economy move forward without this industry being an inside participant? I don't think that's been addressed by administrations over the last twelve years, which encompasses both Democrat and Republican administrations.

Standing together

Many associate our industry's political problems with our diversity and inability to work closely toward common goals. I think the industry is working more closely than in the past. Quite a few of us would like to see this industry speak with one voice. But that may have a touch of Camelot, because we're all pretty strong-willed in this industry. I think that's partly what makes this industry so great. But from the political standpoint, our enemies get to carve us up, divide us, and sink us.

When we stand together, we're fairly strong. As a result, I think you'll see more and more times when the industry can take a coordinated position. We did take a stand on the BTU tax, along with most of our industry. We did take a position on the tax incentives for marginal wells. I think you will find all industry agreeable to taking a singular position on environmental and other regulations, as well, if they are scientifically based and economically justified. I think you're seeing a coalescing of various segments of the industry on some of these major ideas.

I don't think this industry has ever had a leader, per se, after Rockefeller. Hazel O'Leary recently made a statement that this industry needs a leader like Lee Iacocca. Well, I think if you talk to General Motors or Ford they would probably disagree with the fact that Lee Iacocca was ever their leader. The fact is, all companies have fairly strong-willed individuals and I don't know of many times where any industry has gone together and decided that they were going to let one individual be their spokesman for the industry. As long as I'm not a shareholder in your company, I want the company that I'm a shareholder in to stake out its claim and its position and to see what that does for my share of stock.

Reflecting upon this diversity, I think the integrateds, or majors, will continue to play a heavy role domestically, particularly on the downstream side. On the upstream side, the majors have decided to spend more money overseas because of the higher geologic risk of finding anything of size in the U.S., particularly onshore liquids. There will be more and more of a role for what I call the independents. However, I have a real problem with the terms "independents" and "integrateds." I think every company is pretty darn independent. The term "major" has gotten a bad rap because everybody perceives what we call an independent as being one of these entrepreneurial types that really is just out there living day to day and taking huge risks.

We have independent producers and we have independent refiners. Diamond Shamrock says it is an independent refiner. Oryx calls itself the largest independent producer. Well, as I testified before the Senate Finance Committee when they asked about Marathon, I said I work for Marathon Oil Company. We have an independent producing and exploration entity, and an independent marketing and refining entity. In the old days, when we had the depletion allowance, people made money by producing oil, even if the downstream lost money. That no longer exists.

When it comes to capital allocations in a company and when it comes to having exposure on the upstream and downstream sides, both entities have to offer financial return commensurate with the risk, or you'll move capital from one entity to the other. Everybody needs to give a new thought process as to what independents are, and what integrateds are, and what majors are. The only difference between Marathon and Oryx and Marathon and Diamond Shamrock is that we just happen to have both entities under the same name. But I would suggest to you that if you were to go to Exxon nowadays, you would find that their E&P operation is pretty independent from their downstream operation. And they have to be, because if they don't earn a return, the board of directors is probably not going to allocate funds for one of those operations to grow.

This does not imply that we're going to see even greater diversity in the industry. I think what we're seeing is that companies are setting different goals. Those with large volumes of production will have to go where there are opportunities for large reserves, in order to exist and to maintain their income streams. Those with small volumes of production not only don't have to go there, but they probably can't afford to go there. I think you'll see less stratification in the industry than you have in the past. You're seeing independents go overseas now. You didn't see that twenty years ago. There's a lot of players around the world now as opposed to twenty years ago, when you would see only the larger companies going global.

Thus, we'll all continue to have some diversified political agendas, but on significant issues such as the BTU tax, I believe you will see the industry having a more unified voice.

On looking to the future, I am reminded of a series of articles by industry executives which appeared in the *Oil & Gas Journal* in November 1994. Each of these articles addressed the question, "What will working for an oil and gas company be like in the year 2000?" My opinion is that this industry is going to be every bit as vital in the year 2000 as it was when I entered it in 1956. It will be global. You're going to need good people. The evolution, as opposed to revolution, of management suggests that those people are going to have more autonomy. Those people are going to have more training. They're going to have to be brighter and more technically proficient. They're going to have to know how to communicate their ideas.

But the fact is, I don't think the upstream side of the business will ever lose the allure of trying to outguess what nature did. That basically is what you're trying to do. Nature is pretty fascinating. In the downstream side, you have whole areas of the world that really haven't entered the petroleum age yet. All you have to do is look at Asia and look at China, look at India. Those are very vibrant economies and in those economies you're only going to progress with energy. I think that energy will come largely from oil and gas.

The opportunities will be present. We are the technological leader in the world in this industry. Therefore, I believe that as those countries open up, you will find opportunities for equipment manufacturers. There'll be opportunities for good personnel. There'll be opportunities for the technology that we have and they need and for the capital that will come and flow from this economy. So these countries should offer great opportunities.

Of course there will be risks, but there's risk in everything you do in this world. In China, for example, I think people who are patient and who can exercise some good judgment will find that the oil business there fifty years from now may be much, much better than the oil business ever was in the United States.

As for Marathon, we'll continue to be very active in the North Sea and to pursue opportunities there. We've got a project in Russia off Sakhalin Island that we've worked on for four years. That is an opportunity we believe is worth pursuing, even though it's frustrating and people say there is a lot of risk. The fact is that not all is bad about Russia. There are some good things about Russia and they've got some technology. They're going through some growing pains. But I really do believe that if you're willing to suffer through it with them, that there are some great opportunities there.

Southeast Asia poses great opportunities. We're in Indonesia. We're looking at other countries in Southeast Asia. We've always had some operations in Northern Africa. Right now we're in Egypt, Tunisia, Syria, and we were one of the early operators in Libya. We're looking at South America. We had a small operation in Argentina, but I believe South America has a lot of romance to it because governments are trying to privatize. We're really trying to focus our foreign operations in three or four different areas, those being the North Sea, Southeast Asia, Russia, and South America.

Ten years from now, Marathon will probably be more global than we are today. Outside of the North Sea, we're still basically domestic. But if our explorationists are successful and we're smart enough to do what has to be done, I'd like to think that we're going to have income streams from countries that we don't even remotely think about today. And in this country, we'll still be producing oil and gas and we'll still be a major marketer and refiner.

While the search for reserves will continue to drive companies abroad, I think the industry needs to reflect a little bit. I don't want to minimize reserve replacement, but I don't really care as much about reserve replacement as I do income stream replacement and income stream growth. You add reserves in accordance with accounting principles that allow you to put reserves on the books. If you get a service contract with a country, you basically get an income stream, but you may not be able to put any reserves on your books. Even so, you get a very good income stream which is very pleasing to your shareholders.

So my feeling is, if I can get income stream growth, I believe the shareholders will be much happier than with reserve maintenance. But you can't find reserve replacement here in the U.S. Marathon produces the equivalent of about 140 million barrels of oil annually. To think that you could find 140 million equivalent barrels annually in the United States, you'd have to be smoking something. So you have to go overseas. That doesn't mean that you're not going to explore in the U.S. It's just that I have to look at my income stream from the United States, and what I can do to maintain or improve that income stream. I don't care whether it's oil or gas, I'm just looking at the income stream improvement, and the same with overseas. But from a reserve standpoint, you have to go overseas.

A decade from now I see our industry being bigger and better. Nobody's clairvoyant enough to predict the political events. But if you assume, as most forecasters will, that annual demand is going to grow by a million barrels a day, this means that there's going to be an immense amount of capital required to get an extra 10 to 15 million barrels a day in the next ten years.

You've got to recognize that if demand grows a million barrels a day every year, this means you not only have to replace the depletion, you've got to find that new million. This means that over the next ten years this industry is going to have to find and bring on production probably 12 to 15 million barrels a day, or roughly 20 to 25 percent of worldwide production. So there's a lot of growth left in this industry, even if annual demand growth may be only 1 to 1.5 percent.

As I've said, however, there are three broad issues or areas of exposure confronting us as we move forward—technology, capital, and politics. All of them can be manageable, with the political exposure being the highest variable that needs to be overcome. The other two will happen if basic economic law prevails as we go forward.

CONSTANTINE S. NICANDROS

president and CEO of Conoco Inc.

Few individuals have put their personal stamp on the oil industry as indelibly as has Constantine S. Nicandros, the charismatic president and CEO of Conoco Inc. Born in Egypt of Greek parents, Nicandros, known to friends and associates as "Dino," is a graduate of École Des Hautes Études Commerciales in Paris and holds a law degree and a doctorate in economics from the University of Paris. He also has an MBA from the Harvard Graduate School of Business Administration.

Nicandros' interest in the oil industry began during the 1956 Mideast war, when his father was unable to send him money to finish his education. Harvard found some work for him, but "I wasn't eating much," he recalls. Recruiters from Continental Oil Company—Conoco's predecessor—wanted someone who could write on international business in the countries they were involved in and held a reception for Harvard students featuring what Nicandros describes as "the biggest bowl of boiled shrimp I had seen in years. I thought it must be a good company that throws a cocktail party like that."

Nicandros came to Houston when he joined Conoco in 1957 as a research associate in the coordinating and planning department. He was named president of Conoco's worldwide petroleum operations in 1983 and became president and CEO in March 1987.

In 1990, as the industry was still suffering from the fallout of the *Exxon Valdez* incident, Conoco unilaterally unveiled an unprecedented nine-point environmental plan that won it wide public acclaim. It was part of Nicandros' imprimatur on his own company and led to him receiving Clean Houston's 1991 Outstanding Proud Partnership and various other national and international awards for Conoco's efforts to incorporate environmental values into business decisions and practices.

Nicandros is active in civic affairs and is chairman of the board of the Houston Grand Opera. He was Houston's 1989 International Executive of the Year. In conversation, he establishes the essence of a man filled with humor, vision, and a determination to be in charge of his own destiny.

A broad discussion of the U.S. petroleum industry is impossible without relating it to the rest of the world, because whatever happens within the confines of the American petroleum industry often reverberates globally.

This is partly because of the sheer size of the U.S. petroleum market, which is the largest in the world; partly because of the size of the U.S. industry's physical and capital structure and its involvement beyond U.S. borders; and partly because the United States is where the petroleum industry was born and has evolved over twelve decades.

Petroleum industry investment in the U.S. totals a half trillion dollars in wells, refineries, and distribution systems. This is five times the size of the U.S. automobile industry, and larger than the entire nation's air, rail, truck, bus, and water transportation sectors combined. The industry employs 1.5 million people directly and 6 million people indirectly. Its technological prominence is evidenced by the fact that most of the equipment used by the oil and gas industry around the world is made in the United States.

More wells have been drilled and more oil has been produced in the United States than anywhere else in the world. Even today, as social and geologic realities shrink the U.S. industry, it still claims two-thirds of the world's oil producing wells, and volumes great enough to make us the world's third-largest oil producer and second-largest natural gas producer.

Americans are also the world's leading petroleum consumers, partly because of our country's vast expanse and our widely dispersed population, and partly because cheap energy has been the driver for sustained economic growth for over a century. We use one-quarter of the world's oil, and represent a fifth of the world's refining capacity. Transportation relies almost totally on oil, accounting for 65 percent of the total oil consumption in the U.S.

The average American's traditional approach to oil and gas is rather simple. Americans want cheap fuel to drive their cars, to heat and air condition their homes, to run their industries, and to provide their jobs. They want no interference with the easy availability of fuel. If supply is disrupted, for whatever reason, they blame the oil companies. While they prefer their fuel to be of U.S. origin for security reasons, they are not willing to pay anything to encourage the continued existence of a strong domestic industry.

And furthermore, which applies in the last ten to fifteen years with increasing vigor, they believe oil from the source to the pump is likely damaging the environment, and they want it cleaned up, but with no interference to their enjoyment of its benefits. They are somewhat willing to pay for that, but they would prefer the government to provide an indirect way to do so.

It has been and is the role of U.S. oil and gas companies, and the industries that serve them, to meet the public's needs through massive production and logistical capability, enhanced by driving technology, all done in an intensely competitive way. In the sequence of time and history, we first did it in the United States and then throughout the world to meet both the shortfall in the U.S. supply and the rising energy needs outside the U.S. Historically, that endeavor was also a task undertaken by British, Dutch, and French companies.

In the process, major transfer of technology has occurred, resulting in the creation of major entities throughout the world, most of which are government-owned. It is a tremendously diversified industry today, with a great role to play in the development of the world's economies.

So, we as an industry must do our best to come closest to the ideal state required by our customers, which is to provide an unending supply of petroleum (meaning oil and gas), with no impact on the environment, and at a reasonable price.

Now, all of us know supplies of fossil fuels are not unlimited, the price of any product is never reasonable to all people, and our industry by its very nature affects the environment. Yet, our responsibility is to find as many reserves as possible, efficiently enough to make the price as reasonable as possible, thus ensuring that the economic benefit we bring society clearly outweighs any adverse consequences.

This is the theme of man's development, from his first attempts to cultivate the land to the achievements of modern civilization. We strive to develop, yet want to leave a secure inheritance for those who succeed us. And, I believe we have made tremendous progress toward achieving such a goal, and that we are closer than we might think to the ideal state I just described.

Past expectations

To give us a better understanding of our present, it is helpful to review the thinking and expectations that prevailed about two decades ago as to the future of the petroleum industry. In 1973, an unexpected oil embargo had tripled crude oil prices and focused the nation on its supply vulnerability. The perception developed that oil was a scarce and depleting resource. The events of 1978 caused another price shock, and as a result, oil was expected to reach $100 a barrel in the 1990s.

Natural gas was plentiful and everyone enjoyed its use without much further thought. There was some talk about how little new gas was being found, but the persistence of the "bubble" quieted the talk. Coal, with its vast U.S. reserves, was seen as an immediate savior. Nuclear energy was viewed as a growth industry, and uranium was expected to be in short supply. Synfuels were promoted aggressively and research expanded for renewable energy forms such as solar and biomass.

Oil exploration was stepped up at the same time that land availability was circumscribed. Conservation was promoted and gasoline consumption was expected to decline, eliminating any need for refinery expansions.

Fears of an energy shortage stalled a budding environmental movement. Improvements in automobile emission standards were delayed, and power plants were again allowed to use high sulfur fuels. It seemed that environmental protection would have to give way to the preservation of industrial society.

Present circumstances

That briefly sums up the general thinking about the energy future of the United States in the 1970s. It is interesting to note how reality has diverged from what we thought were almost inescapable trends. As seen throughout history, the conventional wisdom of the day failed to consider the ingenuity that would come from sheer necessity as well as some rather basic laws of economics.

We found that, over time, the law of supply and demand works in the oil market just as it does in other markets. In my view, this interaction of supply and demand is a law of nature.

Higher oil prices, coupled with technological advances, spurred the growth of new oil supplies from countries like Norway and many others—both OPEC and non-OPEC—that had been vastly underestimated in terms of their ability to help satisfy a world thirsty for oil.

Since 1973, proved world oil reserves have grown 50 percent to 1 trillion barrels, even though the world has consumed more than 400 billion barrels of oil. Natural gas reserves have more than doubled to 4 quadrillion cubic feet, even though 1.3 quadrillion cubic feet has been used.

After two decades of turmoil, the price of oil has come full circle as supplies have increased. Today it roughly equals—in inflation-adjusted terms—what it was in 1973. The lower price is pushing up demand everywhere, and surprisingly fast.

As all this evolved around the world, where did it leave the U.S.?

These developments had a decided dampening, if not terminal, effect on alternative sources of energy.

Environmental concerns prevented coal from assuming its role as savior, even though it is a strong contributor to the generation of electricity. After a waste of several billion dollars, synfuels have proven to be uneconomic, and will remain uneconomic for the foreseeable future. Nuclear power is deemed an unacceptable risk by a vocal portion of the populace in the wake of Three Mile Island and Chernobyl.

Passive solar energy is making a tiny though worthwhile contribution, but solar energy in general is far too expensive, and no viable technology exists to compete with conventional fuels.

So, today, oil and gas continue to dominate the energy picture outside of electricity generation, and even there we are seeing natural gas take a lot of the growth.

There clearly have been some significant changes. Over the last twenty years, U.S. oil reserves decreased some 30 percent. We found ourselves in a tremendous dilemma, weighing a desire for energy independence with a mounting perception of environmental problems which could arise from efforts to expand supply. By omission or commission, we determined as a nation that we were going to import more oil and not worry about developing domestic reserves for price and environmental reasons.

Today, oil imports fill roughly half our demand, and the most promising areas in the U.S. for oil and gas developing, spanning some 400 million acres, are barred to exploration.

While our oil import dependence is increasing, the prospects are bright for U.S. natural gas. As gas prices strengthen, supplies of this clean-burning fuel will increase, due to stepped-up exploration. We estimate that the potential for reserve additions at prices currently within reach are of the order of 500 to 600 trillion cubic feet. And new, significant investments in transportation and storage facilities are making natural gas more accessible to customers across the country, thus spurring demand.

On the petroleum products side, mainly generated by domestic refineries, there have been tremendous changes. Much cleaner products are being manufactured today than could have been imagined in 1973, because of both regulation and the oil industry putting its mind to it. And I am convinced that the industry can do much more in the future if allowed and encouraged to use its imagination.

In this context, a significant and growing percentage of us in the industry have and are taking steps to make environmental protection an integral part of the way we think and work. At Conoco, we encourage each individual to be an environmentalist. As a result, through a proliferation of new ideas, we have discovered that we could achieve major improvement by releasing the creative energy of a committed workforce.

In 1990, we created a set of voluntary environmental initiatives well beyond the requirements of the law. The initiatives covered emissions and waste reductions, double-hulled tankers, environmental councils in communities, and recycling activities that could be done by every single employee. We also added environmental performance to each manager's compensation criteria.

By these actions, we are making it clear that protection of the environment is just as important as safety, and as the discovery and production of oil and gas.

Industrywide, drilling has become far more environmentally friendly—and cost efficient—with drilling and completion fluids that are increasingly biodegradable. In the past ten years, oil spills from all sources have been reduced to a tiny fraction of the petroleum products used: three-thousandths of 1 percent.

New catalysis technology has revolutionized the refining business. In 1973, we wondered if octane requirements could be met without lead. Today, the FCC [fluid catalytic cracking] process and its catalysts produce more lead-free gasoline at higher octane levels than we ever thought possible twenty years ago.

We are meeting the environmental challenge without sacrificing the growth in living standards. That is the hallmark of development. We are answering the needs of the millions of our customers who depend on our enterprise and expertise to deliver those solutions. With cleaner fuels and better cars, hydrocarbon tailpipe emissions in the United States have been reduced by 96 percent since 1960.

Our refineries have been retooled so that their emissions represent only 2 percent of all industry releases. During summer, the vapor pressure of all gasoline has been reduced in order to lower ozone-forming emissions. All diesel for highway use is now low in sulfur. In short, the air in the United States is cleaner than it ever has been since the industrialization of America.

In the past twenty years, gasoline demand has grown by almost 10 percent, even though our automobile and truck fleet is 40 percent more efficient than it was then. The reason is that we are driving more miles—75 percent more than we were in the early '70s.

Even though we can see amazing progress from the vantage point of hindsight, we must not overlook the turmoil and real pain of this industry caught in the whirlwind of change. A half-dozen major oil companies disappeared through mergers. Seven thousand independent companies folded. The oil service sector was decimated. Two million jobs vanished in many related professions, causing immeasurable hurt and heartache. The net impact has been a much smaller U.S. domestic industry than that of twenty years ago.

Capital flight from the United States became a fact of life as the surviving companies shifted billions of exploration and development dollars to more hospitable shores. The number of refineries shrank by 28 percent, but those remaining became far more sophisticated, expanding capacity by 12 percent.

All these events represent a natural evolution in a changing world—either you change with it, or you go by the way of the tide.

Future progress

So, what about the future? While the price of oil today and rising laws and regulations—environmental and other—could cause us to be a bit despondent about the future, by giving in to depression we would be underestimating adjustments that can and will be made. Even though we are on the flip side of the cycle, the same ingenuity that got us through before will get us through again.

Society depends on us to provide an unending supply of energy with no impact on the environment and at a reasonable price. And we are remarkably close to doing it, and in all likelihood will continue to close the gap between the real and the ideal.

Let me first address the issue of cost. The world marketplace has been extremely effective in keeping prices low—perhaps too effective from the point of view of producers, but not so from the point of view of our customers. Most forecasts call for continued low prices through this decade. This satisfies one of society's demands. At the same time, it makes another demand much more difficult to meet—that is, to ensure plentiful supplies while having less revenue to find and produce them.

Technology continues to be our best ally to push the frontiers for profitable developments in a low-price environment. But let me emphasize, in case there are some people who think otherwise, that technology's rose is not to build scientific monuments, but to ensure that we can find, develop, produce, and operate at a cost that will give us adequate return on investment under current price conditions. And technology is helping us do that. Advanced exploration tools are enabling us to reduce the cost of finding new oil and gas reserves. New 3-D seismic processing procedures improve the clarity of subsalt images.

Interpretation techniques such as sequence stratigraphy help us correlate sediments on both geological and geophysical data so that we can more accurately spot potential reservoirs in a basin. We are replacing traditional materials like steel-coiled tubing with exotic composites that are lightweight, corrosion-resistant, and offer advantages in deep water and Arctic offshore environments.

Low oil prices are causing technological advance to be pursued differently than in the past. Stand-alone research efforts are being replaced by strategic alliances within the industry.

We are also benefiting from the peace dividend. The U.S. government has swung open some doors of its defense labs to private industry. To mention but one instance, Conoco, together with other companies, has joined government scientists in an effort to build a downhole seismic source to allow us to more accurately characterize reservoirs between wells. Each company has contributed its prototype tools, and government scientists and engineers are making enhancements through the application of missile science.

Another advancement we in Conoco are excited about is the evolution of our own deepwater tension leg platform [TLP]. While a third generation of this technology, this time with concrete, is being applied in the Heidrun field in Norway, we are testing a fourth generation—a micro TLP system. It can be moved and reused in stages to develop clusters of small reservoirs, greatly reducing costs of deepwater developments.

To be faithful to fulfilling the need for endless supplies, we have to keep pushing the envelope until we find the true limits of our expertise. And, I can safely say we are nowhere near that point today—if, in fact, it exists.

Now what about the environment? As the issues of supply and price have receded, U.S. policies—like those of other industrialized countries—have increasingly centered on the environmental effects of petroleum's use, and the industry is responding.

Here, too, we continue to meet new standards while striving to better them in the future. On the first day of 1995, U.S. refiners began producing cleaner gasolines to cut automobile emissions by another 15 percent.

The massive, multidisciplinary, multimillion-dollar joint research effort between U.S. automakers and petroleum companies—which provided much of the data used to design the cleaner fuels—continues to test and analyze various combinations of vehicles and fuels so that this progress can continue into the next century.

During the next decade, our refineries, natural gas processing plants, and all distribution facilities will be working to meet the objective of installing "maximum achievable control technology" to further combat air pollution.

Obviously, larger environmental issues—like global warming—also deserve attention. I will not pretend to even outline the scientific debate that surrounds this issue.

The wise thing to do is give science a chance to prove over time what needs to be done. Political correctness and expediency are overtaking this subject, and even though this is well meaning, we should think carefully about the consequences of all we do. Certainly, further study is warranted before drastic steps are taken that could send economic growth in a tailspin and affect the livelihood of millions of people.

In any event, a sensible conservation ethic around our use of energy makes good sense, regardless of what the science says today. We must continue to search for efficiencies to reduce consumption in industry, automobiles, and other transportation equipment.

Caveats

So our industry can be proud of the progress we have made, and are making, on the triple front of supply, price, and environment. As we go even further, we must do so prudently, with full awareness that changing circumstances could force us to alter our direction.

As an oil and gas executive, I worry about oil prices that go down, about taxes that go up, about my government and others that do not appreciate that lower prices and higher taxes can be lethal to industry and jobs. I worry about the call for investments that do not produce jobs, about proliferating regulations, and so on. In 1992, the U.S. petroleum industry spent $10.5 billion on the environment.

This is equal to the profits of the top three hundred oil and gas companies that same year, and is more than the entire industry spent on exploration in the United States. As our costs rise under the pressure of a myriad of regulations and the price of our products does not, I worry how we can continue to succeed and raise the enormous capital needed to meet the world's demand.

As a human being, I worry about the state of my country, and for that matter the planet that my grandchildren will inhabit. I want to do my part in seeing that the environment is protected, and I want humankind to continue to improve its economic lot. After all, poverty is at least as undesirable as pollution. I worry whether we are assessing carefully the consequences of all environmental measures to avoid committing false environmental economies.

Do we reduce emissions in one place, but raise them somewhere else—such as in the process of generating power for electric cars? Do we look at every action in the context of other actions we take?

Finally, I worry about governments or ideologies that want to manage my life and that of my family from birth to our final days.

Despite these worries, I am convinced that our industry is capable of performing to society's satisfaction, although the standards demanded by society continue to spiral ever upward. Some have called our industry a sunset industry. To them, I submit that in our development and application of technology and in our responsiveness to a changing operating environment, we are second to none.

I remind everyone that we are the industry the world depends on—and will depend on for a very long time—for practical, affordable, reliable fuels that give us heat, light, transport, manufactured goods, and benefits we can no longer imagine living without.

Supplies today are plentiful. While the United States has by no means solved all its energy problems, we are increasingly confident that we are more capable of managing supply disruptions than in the past. Restructuring and technology have improved the way we do business and have helped drive down costs.

While certain parts of society will never be satisfied with the environmental headway we are making, responsible people will support a balanced cost-benefit-based approach to the equation linking environmental improvement and economic growth.

We will continue to reconcile the use of petroleum and society's environmental imperative. Producers and consumers have a mutual interest in the technological strides needed to keep petroleum competitive—environmentally and economically.

If we look at the oil and gas sector as it approaches the millennium, we see a 136-year-old industry which has time and again confounded the prophets of doom, which has equipped itself with the technology and the management approach to deal with whatever challenges a changing global environment throws at it, and which will enter the twenty-first century with record reserves and every prospect of increasing those reserves for the foreseeable future.

We, as an industry, can be proud of our increasing sustainability and our achievements, which serve the demands of our customers, our shareholders, governments, and society at large.

R. E. (RAY) GALVIN

president of Chevron U.S.A. Production Company

R. E. (Ray) Galvin is president of Chevron U.S.A. Production Company and a vice president of Chevron Corporation, originally Standard Oil of California.

Galvin is a native Texan who holds a B.S. degree in petroleum engineering from Texas A&M University. He began his career with Gulf Oil in 1953 as an engineering trainee and worked on a number of assignments in Oklahoma, Pennsylvania, Texas, and Louisiana.

Galvin became Gulf Oil's vice president of production, U.S. operations, in 1979 and was named vice president of the company's South and East offshore division in 1981. He was elected vice president of exploration, land, and production for Chevron U.S.A. Inc.-Southern region, in the June 1985 merger of

Chevron and Gulf. In August 1988, he was elected senior vice president, exploration, land, and production for Chevron U.S.A. Inc., formerly the domestic oil and gas subsidiary of Chevron Corporation. He was elected to his current position on January 1, 1992, when Chevron U.S.A. was split into two units, one for exploration and production, the other for refining and marketing.

Galvin, who was interviewed in November 1994, is a highly visible leader and spokesperson for the petroleum industry. He is chairman of the Natural Gas Supply Association, vice chairman of the Natural Gas Council, on the executive committee of the Mid-Continent Oil and Gas Association, on the executive committee and the board of the National Ocean Industries Association, and on the board of the Gas Research Institute.

I look at the industry from a domestic upstream perspective, and one of our vital issues is the fact that for twenty years we've had declining oil production. There is a question as to whether natural gas production will decline. Gas production also peaked in the early '70s, declined some, and we're now at a second peak.

The question is, Can we as producers sustain that increase? If we can't, we're in the terrible predicament of being in a commodity business with declining production. This brings all sorts of price-cost pressures that cause us to need to continually improve our productivity in order to compete on a flat-price basis. Certainly it's been a flat-price basis for crude for years with a few blips, but now it also appears there may not be much real growth in natural gas prices. And that's the problem: to be in a declining volume business with flat prices and fierce competition.

I continue to believe that it is important for this country to have a strong domestic industry, but I've come to realize that it may not be as critical as I once thought it might be. I think the fact that international oil production is no longer concentrated in a couple of countries gives protection we may not have had years ago.

I have come to this conclusion from personal acceptance that our domestic oil production is going to decline no matter what we do. So we must figure out a way to live and prosper in a world where we're not self-sufficient in crude oil production. I think one of the ways you do that, though, is to continue to have a domestic industry that is as healthy as it can be. But I'm not sure that our nation can afford to subsidize our industry, to any substantial extent, for us to be healthy.

I think government needs to do the things that they can do, such as pass reasonable environmental laws that don't prescribe specific antidotes but lay out the goals and let us use our technology and business skills to meet those goals. Government should also provide a level playing field between forms of energy. If government does these things, then it's incumbent upon us as operators and practitioners in the industry to make ourselves healthy. And sometimes making ourselves healthy means that we have to take some pretty tough medicine.

Remaining competitive

The thing that immediately comes to mind is the downsizing that's going on, which is the dark side of improving our productivity. Obviously, we have to get more productivity out of fewer people. People are one of our larger costs. Our production continues to decline and, necessarily, the number of people that we have is going to decline also.

More gratifying is the fact that the effective application of technology is really the only way we're going to be competitive and continue to have a viable and healthy domestic industry. As prices and the expectations for real price growth have deteriorated, the tremendous gains that we've made in technology and the application of technology over the last six or seven years have really been the bright spots in our industry.

We've accomplished some great things. I believe it was [deputy energy secretary] Bill White who said at a meeting that we're in a race between technology and depletion. When we as an industry don't have expectations of real growth in prices to help bail us out, it causes us to really focus on the things that we can control. One of these is the application of technology.

We need technology more than ever. If you look back, there has probably been a faster rate of technological development in the industry during the times when prices were poor, the '50s, '60s, and '70s, than during the "boom days" when oil prices increased so rapidly that no one worried about improving technology. The only questions then were how fast could rigs be built and wells be drilled.

I believe we'll actually see an improvement in the rate of technology development. We need it to be cheaper, and we need it to be focused on things which can help us right now. We can't spend a lot of money on enhanced recovery processes that wouldn't be economical unless oil got to be $35 or $40 a barrel. But anything that will help us get more oil out of the ground at $20 a barrel—that we need. We need technology that could help us drill more cheaply. We'll gladly pay for a process that will help us fracture treat better.

The improvement in computer powers and the attendant lowered cost of geophysical computing lets us accomplish seeming miracles with 3-D seismic. This is what's letting us economically find the smaller new fields and the additional undrained oil and gas in the nooks and crannies of existing fields.

The manpower dilemma

A real problem that we face now concerns the hiring of graduating petroleum engineers and geologists. We need to continue to have them coming out of the universities. We'll need them in our company 5, 10, 20 years from now. But it's very difficult to be out hiring them at the same time we're downsizing.

We're aware that we need to get that new blood in, so we're bringing some of it in. We are going for the middle ground. We are hiring a few to fill some specific needs from year to year, but we still may not be getting as many as we will need to have in that age group. Ten years from now, my successors will probably look back and ask why we didn't set them up better.

On a comparative basis, in my part of the company, which is domestic exploration and production, where we had 8,300 at the end of 1991, we'll probably end 1995 with about 5,200. Most of that reduction occurred in 1992, but in 1993 and 1994 we reduced between 4 percent and 5 percent per year. We will probably match that performance in 1995.

And that's attributable to trying to get ourselves in shape to be competitive. You need to compare the number of people that we have running the business with what we had before the big oil boom of the late '70s and early '80s. We, as an industry, are just now beginning to get back to the same level of employees per well or per barrel that we had then.

True, the complexity of the jobs we perform has increased. We have a lot more environmental regulations. We have more wells producing less oil. But we developed some very bad habits when we thought that oil prices and gas prices were going to continue to increase rapidly. We developed a lot of unnecessary overhead that we're still working our way out of.

We've retrained and redeployed a number of people. All of our personnel reduction didn't leave Chevron Corporation. We were able to redeploy some four hundred in other parts of Chevron. We have geologists working in chemical plants and facilities engineers marketing lubricants. We have done a reasonable job of redeploying and retraining the people who were interested in changing.

A number of the employees preferred not to avail themselves of those opportunities. They wanted to go outside and attempt to find a job in their current career field. Some of them have been very successful. Others, unfortunately, have found that it's a pretty tough world out there, and that there's an oversupply of their skills in our industry.

The majors are going to have to continue to change, but the changes may be continuations of the trends of recent years. The Lower 48 oil production peaked in 1970 and it's been declining since. It's probably going to continue to decline, so we're going to have to continue to adjust our employment levels.

Partnering

A number of the independent producers have performed very well in the last few years by finding niches. Many of these have established a position in the natural gas business. Others have purchased producing properties from major companies and set themselves up in specific areas where they can be a low-cost competitive operator.

We've always partnered with independents. In the old days it had to do more with exploration farm-outs on smaller or marginal prospects.

Where we're having some of our most interesting contacts and joint efforts with independents now is in nonconventional gas: tight sands and shale gas. We set up a group in 1993 to obtain a bigger position in tight gas sands or shale gas areas. In several of these areas there weren't many other majors.

Our people would go in and try to find a small producer using "best practices" in that area. In several of these locations some very small companies were the ones using what we perceived to be best practices. They were the ones using technology from Gas Research Institute, or using operating practices that just made sense in their specific location. They knew the territory. Many times they lived right there. In many cases, they were cash-flow constrained and we were able to work out some joint endeavors that made sense for both of us. So we're learning a lot from them.

I think overall there is more of a mutual respect now between us. I see that in our industry organizations, although I still see some division. And maybe that's as it should be. There are certain issues that we may not ever get together on, but there are so many issues that we should agree on that it's good that we do now have a better relationship. The ranks of the independents have thinned considerably. The ones who have survived are obviously good business people and are doing many things right, because it's been a tough world.

It's up to us to figure out how to be healthy. And healthy is going to involve us majors learning a lot of the tricks from independents, including working jointly with them in places where they can add value. There are certainly a lot of places where they can. And "healthy" would include us continuing to use the special skills and resources that we have as major companies to make money for our shareholders.

Suggestion to government

An energy policy should be logical, not political. Unfortunately, it just doesn't work that way. Because of the quantity of our imports, policy makers need to understand that despite all of the encouragement they can give to go to alternate fuels, this country will still need all the domestic oil and gas that we can produce. They need to understand that there's room for every barrel that we can economically produce—that this production would only reduce imports, not stifle alternate fuels.

I believe energy policies should not favor any energy source over another. Environmental policies should set forth standards and various energy sources should compete based on their ability to economically meet these standards.

Given the tremendous reserves of oil in the world, we're probably another forty to sixty years before other forms of energy seriously challenge oil. Petroleum companies are involved in coal, geothermal, uranium, and solar. Some have been successful. We characterize ourselves as energy companies and if it's a form of energy that makes sense, I would believe at least part of our industry is going to be involved in it. However, I do think some companies have decided they might do best by focusing on their core oil and gas business.

I think it's understandable for some parts of our industry to believe that there ought to be an import fee. I've personally come to the point that I don't agree, but I think it's appropriate for people to express and advocate their opinions about that. I guess I've seen that every time government has given us something, the process has gotten incredibly complex, and they've ended up taking away more than they gave. So I'm somewhat cynical about getting anything that will work.

There are probably some areas where tax incentives would encourage development of opportunities that would not come about otherwise. It's hard to design them to accomplish only the intended purpose without affecting other things. I think the Section 29 credits for coal-bed methane and tight gas sands encouraged a lot of development that might have been a long time coming otherwise. But they ended up generating a windfall for some companies far beyond what was necessary to generate the desired level of activity. Incentives, if they come, should be carefully designed to provide the intended result and have limitations that would keep them from going beyond that.

Lawmakers should—not just with our industry, but with all of us— limit their intervention into establishing rules that set out the desired result and don't tell you how to get to that result. They should not establish rules that have particular political bias that enrich one portion of the nation at the expense of another. That's just equity.

More access

A sound energy policy would also assure that access to prospective areas onshore or offshore would not be denied without good reason. Such decisions would be based on sound science, not political hysteria.

Access is a very tough political problem and we have faced it just about everywhere it could be faced in this country. I don't know what the answer is.

The parts of the country that are available to explore have already been thoroughly explored, and most of the areas of the United States that are capable of having large new fields discovered are off-limits for environmental reasons. That would include areas like offshore North Carolina, offshore southwest Florida, Bristol Bay in Alaska, and offshore anywhere on the West Coast. Most of these areas are in moratoriums until the year 2000. Some of them may be in moratoriums forever.

So that leaves us with the extreme deepwater Gulf of Mexico off of Louisiana and Texas, the subsalt, and some very rank onshore prospects, usually very deep. And you have to have something large to make projects economic in the deep water, or else we must have real growth in crude prices. Most companies' expectations concerning real growth in crude prices are that it's going to be pretty slow for some period of time.

I would agree there are a few places in the country that—although the risks of an incident are small—resulting damage could be so great that I wouldn't choose to operate in them. But those are very few.

We've demonstrated by our operations over the last twenty to twenty-five years that we can appropriately operate in most locations, including sensitive offshore locations. Our problem is that the public's evaluation of risk is so skewed by the information that's fed them by some of the environmental organizations, primarily to support their fund-raising appeals, that we don't yet have good access laws. We don't have good environmental rules. Indeed, we have such things as the people of Florida continuing to import almost all of their oil through tankers and barges.

They accept this despite spillages of oil from tankers over the last twenty to thirty years that are far beyond what you'd statistically expect from a producing operation. There is tanker traffic near the Florida Keys and the Florida coast year after year and yet Florida is adamantly opposed to oil or gas production operations within one hundred miles of its shoreline.

We just drilled a well in Destin Dome after a three-and-a-half-year struggle to get permits. This particular wildcat was not productive, but we do have a productive field offshore Florida that we will be proceeding to permit for production.

We have some exploration leases off the west coast of Florida in the Pulley Ridge area that we'd like to drill. We have leases offshore North Carolina; we'd like to drill there. But we don't want to spend the money on exploration unless we're going to have the ability to produce something if we find it.[1]

Note of optimism

An article in *Geophysics*[2] magazine by Dr. William Fisher of the University of Texas is quite optimistic. He said there are 150,000 fields left to find here in the Lower 48. The bad news is that about 135,000 of them have less than a half million barrels of reserves; they're from 50,000 to 500,000 barrels. You're not going to make a whole lot of money finding those kind unless you find them right where you already have infrastructure, unless you find one nine times out of ten when you drill, or unless you find very many in a small area.

1. On July 31, 1995, the Clinton administration agreed to repay nearly $200 million to nine oil companies that had paid for the right to drill off Florida and Alaska but then had been prohibited by drilling bans. The settlement covered seventy-three leases off southwest Florida and twenty-three leases in Bristol Bay, Alaska. Of the companies, Chevron is to receive $65 million and Conoco $23 million. The companies had originally sued the federal goverment for $1 billion after the Bush administration and Congress imposed a drilling moratorium covering most coastal waters.
2. W. L. Fisher, "Future Supply Potential of U. S. Oil and Natural Gas," *Geophysics: The Leading Edge*, December 1991.

Only a small percentage of them will be the size fields that have historically been economic for a large company. That's the problem that we as majors face. We have always worked best when we could take advantage of economies of scale, when we found something very large which could carry our overhead. Now the production on those large, older fields is dwindling to the point where we must be sure we are shrinking our overhead to match the shrinkage in production volume.

We must use ever-improved technology and we must apply that on a cost-effective basis. We must adjust to the decline, adjust to the fact that the "factory" we're working with is a hole in the ground that may have been drilled sixty or seventy years ago. A very aging resource. Our challenge is to roll down operating costs, protect the environment, and get more oil or gas out of that reservoir and/or find some other reservoirs close to it.

At Chevron, our capital program for the next year or two is still roughly fifty-fifty oil and gas. But our product mix is moving more and more toward gas. According to most studies of remaining resources, there should be more opportunity to find gas in the U.S. than there will be to find oil.

Top priority

Environmental issues affect almost all facets of planning. Access issues affect exploration and production—whether we can do it at all or whether we can expect inordinate delays and higher costs. This is a major factor in determining where to spend exploration money. Likewise, the threat of increased regulation of existing producing operations affects the allocation of capital between projects, and even between this country and other countries.

A serious environmental incident could cost several years' profits. That's the bad side. The good side is that the public seems to recognize when you do a good job environmentally. From a marketing standpoint that's a positive, and it's also a positive when you're attempting to go in and produce in new areas. We take people from areas where we hope to operate out to see the operations that we already have, because we're proud of what we do.

All of our people have not only the authority, they have the responsibility, to shut down not only any operation that creates a violation of the rules and regulations, but also any operation where they think an unacceptable potential exists for an environmental incident. On a recent wildcat well near Jackson, Wyoming, we hauled the rig and all large loads in by helicopter because we felt that was less environmentally sensitive than building a road through one area susceptible to landslides. On that particular well, we cleaned out two existing old dumps that had been there, probably for over fifty years. This was a way that we could leave the area better.

Our country's population is continuing to grow. We must accommodate that population with a decent living standard. They don't want to revert to no electric lights, no air conditioning, and no cars. They want enough food and they want good food. But people want—and deserve—a clean environment. We must recognize that this doesn't have to be an either/or situation—either development of oil and gas *or* a clean environment.

We can have both. If we're really going to succeed as a nation, we must have both. We must get responsible scientists, responsible producers, and responsible environmental organizations to work together to achieve that goal. So far, I have only seen faint glimmerings of hope that this will occur.

Natural change

If we could get a message across, it would be to look at us as we are and not as a stereotype. The last, and really the only major, offshore oil spill from an exploration and production operation in the United States occurred in 1969 [Santa Barbara]. The procedures and processes that we use now are no more like those used in 1969 than are the cars we drive, the airplanes we fly, or the environmental rules.

We're capable of changing; we've recognized the need to change and we're capable of operating by rules that are consistent with our understanding of the need to protect the environment. Chevron conducts public opinion surveys periodically in which we ask people how they view this oil company or that oil company, and then we ask, "How do you view the majors as a group?" In recent years, with only one brief exception, every major oil company *viewed separately* scored higher than the industry as a whole.

I doubt we can change this aspect of human nature that causes people to tend to resent entities upon which they are very dependent, particularly if this is coupled with a feeling that they've been taken advantage of. You can look at the way the press handles reaction to increases in the price of gasoline. Everybody's buying natural gas or heating oil for their furnace or certainly gasoline, so they're all affected when we do something. There are not too many other industries that affect so many people so quickly if prices rise.

The rest of this decade will be very interesting and challenging years. Can we get the public to accept the need to continue to explore for and develop domestic energy sources? Can we economically find and produce what in most cases will be smaller and smaller targets? Can we continue to improve technology rapidly enough to compete with countries whose wells produce 10,000 barrels a day when ours produce less than 10?

With this kind of an outlook for our industry, can we continue to attract and hold some of the brightest and the best of our young people to develop and apply this technology for us? Can we form that ultimate alliance—an alliance with responsible environmentalists who would work to find places that we could agree would be appropriate for additional responsible exploration and development?

W. W. (WAYNE) ALLEN

chairman and CEO of Phillips Petroleum Company

W. W. (Wayne) Allen is chairman and chief executive officer of Phillips Petroleum Company.

Phillips is one of the nation's best-known integrated petroleum companies and is involved in worldwide petroleum exploration and development. It is a leading refiner, gas processor, and producer of natural gas liquids and is also involved in the manufacturing and marketing of chemicals and plastics. In November 1969 Phillips made the first major discovery in the North Sea, with a find in the Ekofisk field near Norway.[1] The company is also a leader in subsalt drilling in the Gulf of Mexico.

A native of Arkansas, Allen has a B.S. degree in mechanical engineering and a master's in industrial engineering from Oklahoma State University. He was inducted into the OSU College of Engineering, Architecture, and Technology Hall of Fame in 1992.

Allen joined Phillips as an engineer in 1961. He became director of drilling and production for the Europe-Africa division in 1980 and became operations manager of the Ivory Coast region the following year. In 1984 Allen was named United Kingdom regional manager in charge of exploration and production. He became general manager of Western exploration and production in 1986 and two years later was appointed vice president of international exploration and production. He served as senior vice president of exploration and production from 1989 to 1991, when he was named president and chief operating officer. He attained his current position, replacing the retired C. J. Silas, in 1994.

In this interview, conducted in April 1995, Allen gives a clear picture of an industry and a company in transition. He begins by going directly to the heart of the industry's dilemma: the lack of access to attractive drilling prospects.

*I*n the upstream, domestically, we have a fundamental, long-term problem of access to prospective acreage. And it's having a pervasive impact on the industry. Companies are having to come to grips with the grim reality that many of the best areas—from the Arctic National Wildlife Refuge to the Outer Continental Shelf—are off limits.

Phillips' production in the United States is under pressure, as we're feeling the impact of limited access to high-priority exploration areas. Our worldwide oil production is increasing, yet it's falling here at home. Our experience is typical of what the industry is facing. Ideally, we'd like to see production increase at home, where the tax structure is more favorable. This is one reason our success in the subsalt oil play in the Gulf of Mexico is so important.

1. According to Daniel Yergin in his book *The Prize*, Phillips made its discovery after reluctantly deciding to drill one more well because it could not find anyone else willing to sublease its rig. Ironically, experts recall that ARCO also discovered oil on the North Slope in Alaska after it, too, reluctantly decided to drill one more well.

With so many U.S. properties off limits to oil and gas exploration, the industry is responding by redistributing U.S. properties—buying and selling assets—and by shifting its exploration emphasis overseas. Where there have been tax incentives domestically—such as in deep drilling or coal seam gas—the industry has responded.

Typically, environmental issues have led to limited exploration opportunities in the United States. The exception to that has been California, where even with federal approval, the industry has been blocked by local governments from carrying out oil and gas activities. There has to be some way, in these cases, where the industry has the assurance it can move ahead.

It's happening overseas. The international contracts with the host governments are very well spelled out. So you see the major oil companies, and large independents, starting new ventures all over the world—particularly in the former Soviet Union, China, Eastern Europe, and Africa.

The U.S. gas market also has been quite challenging. The broad deregulation taking place has had a positive impact. And hopefully, deregulation will go all the way from the wellhead to the burner tip. We're not quite there. It's been a plus for consumers and producers alike. And we're seeing a shift from oil to natural gas, in part because of the environmental benefits of burning gas. The companies that have adjusted to this shift through technology, and by keeping their per-unit costs low, are benefiting.

Also, I think we're in the first phase of an important trend concerning the development of major gas resources around the world. The accumulations in the Middle East, particularly in the Persian Gulf area, have been known for many years. Moderate levels of gas production from the old consortium fields are going through Iran into the pipeline in Europe. It seems possible down the road that some of the gas from the Middle East will go all the way to the United Kingdom.

At the same time, the Norwegians are particularly blessed with resources. They've had European gas markets pretty well to themselves. However, it's starting to get much more competitive for Norwegian gas, and ultimately for British gas, in the European markets. Coming the other way, the Algerian link is moving gas to Italy and Spain. If politics stabilize a bit, you could see even more.

Lastly, OPEC continues to need a certain rate of oil production and is in some ways a stabilizing factor in the market. OPEC also can be a large consumer of capital to develop additional capacities. I see commodity prices for oil and gas in the foreseeable future extended out quite a ways. There can be disruptions, no doubt, but overall, the situation appears fairly stable.

Downstream

Concerning the downstream, the United States has many agendas, most of them political, to the point where it's no longer a free market. It's sad that a free society like ours is allocating markets, whether it's for methanol or reformulated gasoline. The mandates are poorly thought out. And the industry has spent huge amounts of capital to respond to this sort of fictional unknown fuel of the future. It's really regressive. And I don't see it getting any better.

Also, we have foreign players now in the market, and it's quite different from anything we've seen before. We have Petróleos de Venezuela[2] owning refineries and marketing through Citgo. And then there's the Saudis working through their Star arrangements with Texaco. Because these foreign governments control the crude oil—and the price—it enables them to some extent to set the market price and to be uniquely competitive. In short, the refining and marketing business in the United States is brutal, and I don't see that changing.

Then, the present administration has a philosophy of backing out of oil by 20 or 30 percent by the year 2000. That's not realistic. It's not in the economic interest of the country and I don't think it's going to happen. Still, it's a political force to be reckoned with. Very challenging. In general, most foreign governments are behind the United States on its fuel requirements, but they're moving in the direction of reformulated

2. The Venezuelan state oil company is referred to by the acronym PdVSA. Its U.S. arm, PDV America, has assets of more than $1.75 billion and is the fourth-largest retailer of gasoline in the U.S. through Citgo.

gasoline and low-sulfur diesel. The environmental movement is strong worldwide, and the factual data on such things as ozone will continue to be hotly debated.

As for the rest of our downstream businesses—mainly plastics and chemicals—I think we'll continue to see a cyclic situation. But because of our focused-growth strategy, we'll have very high growth, specifically in ethylene and polyethylene. We also have a strong position in specialty chemicals. And as more countries modernize, the demand for plastics and chemicals will continue to grow. I'm thinking particularly of nations such as China, India, and Malaysia.

Well, that's the way I see things. For the long run, hydrocarbons will be the fuel base. Coal in the United States has come on, particularly Western coal. Nuclear power will continue to struggle until the public has more confidence that this is an abundant and safe, clean-burning energy source. But there's no magic bullet out there. Oil and gas will shoulder a good part of the burden for many, many years. As for supply and price, I think there are adequate supplies that, barring catastrophic events, can be delivered at bargain prices well into the future.

Our industry has made significant investments in our refineries that enable them to handle sour, heavier, less desirable—and cheaper—crudes in the United States. Now, if oil prices were for some reason to hit $40 a barrel, then you'd see a significant switch. But I don't think it's in the cards.

Technology has been very important in revitalizing the industry. Phillips' pioneering efforts in 3-D seismic interpretation are enabling us to carry out the first subsalt oil development in the Gulf of Mexico—the Mahogany field. But equally important has been our emphasis on using technology to operate better—perfecting the horizontal drilling technique, getting deck weights down, moving into deeper water, improving subsea wellheads.

Perhaps the biggest impact of technology relates to prices. Our industry can no longer count on price increases in general to boost revenues and earnings. Technology helps us break into increasingly more difficult environments—like the deepwater gulf or West of Shetlands—yet keep our costs under control.

Adversarial relationship

The role of government is both a controversial and a complex issue. In air quality, for example, I believe the government should set the standard and then step back and let the companies apply the best technology to achieve whatever the standard might be, rather than by allocating fuel markets.

Many foreign governments see energy as critical to industrial development and therefore work very closely with industry. We never quite seem to get to that point in the United States. The U.S. situation is more adversarial. Our industry has come to accept this. And maybe even our citizens prefer it this way, since they're skeptical when government and industry move too close to one another. But it's quite a different situation here than in many foreign countries.

The 1994 election of a Republican Congress eroded the influence of those who would impose even more stringent regulation on the industry. In my view, the tide had already begun to turn against those favoring excessive regulation, but the '94 elections should hasten that trend. A sizable number of new members of Congress have business backgrounds. That's welcome news, too.

Our company has long taken a pro-active approach to environmental issues. We address environmental concerns in the conceptual stage of projects. And we try to correct environmental problems at existing facilities before mandates are enacted. We think that's doing things the right way. We accept the responsibility to provide safe and healthy work environments, and we realize that we serve many stakeholders—from shareholders to next-door neighbors. We also want to provide jobs in the communities where we do business, and that means keeping your costs down and having a good reputation in these communities.

In general, I'd say the industry's extraordinary efforts are rarely publicized and almost never appreciated. Even so, providing a safe, healthy work environment is essential to our people. It's consistent with our values as a company. Beyond that, our safety record is an indicator of how effectively we're managing our assets. Safe plants are typically more productive and profitable. A good safety record cuts down on insurance premiums as well, which for a company our size can run into the tens of millions of dollars.

There's no question that restructuring will be an ongoing process. With commodity prices often held in check, and with the pressure to keep per-unit costs low, restructuring is no longer a description of an event but a process. This is especially true in the downstream businesses. Companies like Phillips are focusing on their core assets and running them at maximum capacity and efficiency. We're continuing to get out of businesses—or marketing territories—where we can't be a top competitor. Computing technology continues to move forward. And it's reduced the need for much of our clerical help.

Keeping costs down

During the recession a few years back, we at Phillips found ourselves in an unusual and uncomfortable spot. All of our businesses went into reverse gear at the same time. So we made adjustments in our cost structure based on what was essentially a worst-case scenario. Any improvement from that point would be a plus for profits. At the same time, we were able to sell enough non-core assets to help us carry out an aggressive capital program. We knew prices and profit margins would eventually get better, and we wanted to be well positioned when that happened. It's still early in the cycle for some of our big projects, but it's clear that a recovery is underway.

The most important thing in holding down costs is instilling a cost ethic in our people and incorporating that cost ethic into everything we do. One security analyst said we're developing a "profit-driven culture." In the past, we tended to see cost control in terms of periodic campaigns involving significant job losses, rather than as an ongoing discipline. Adjusting the size of our workforce is only part of the process. We also have to make sure we have the right equipment and training, and are using the latest technology and business practices in order to be as efficient as possible.

We can never afford to relax on costs. In a commodity business—like oil or gas or basic petrochemicals—the low-cost producer wins. If we're going to make good on our commitment to deliver superior returns, we have no choice but to be a leader on costs. And we aren't alone. With very little inflation these days, businesses everywhere are being driven as much by costs as by revenues.

The biggest challenge for us is living within our means. We haven't generated enough cash from operations to fund our capital program and dividends for the last five years. With the economy back on its feet and new production volumes coming on stream, we think there's a good chance of closing the gap this year. However, to be truly successful, we have to get to the point where our operations can provide enough cash to cover spending year after year.

For companies operating in the international arena, alliances are a risk-spreading device. Some large major oil companies may have traditionally operated autonomously, while smaller companies were looking at ways to spread the risk. So they took the lead in forming joint ventures to share the technical burden and the costs. Now, it seems, most companies are sharing information more freely, and going to the team concept to hold down costs, move projects to completion more quickly, and remain competitive.

Because there's such extreme pressure to keep costs lower, it's a healthy trend, I think, to see companies pair up to optimize the use of assets. This doesn't mean there's no such thing as proprietary technology anymore. But it does mean we're finding ways to pool our strengths to be more cost-effective—and more responsive to the needs of our customers.

The energy industry is changing. So you've got to have someone who can cope with the changing environment, and who can work with the modern workforce, which is more team-oriented than functionally oriented. You need someone with vision who can inspire a whole workforce to share that vision.

The business we're in today is going to be different than the one we'll be in tomorrow. The leader who has a vision of tomorrow's business, and who can lead a lean, capable workforce into that business environment, with the tools to prosper in it, is the one who'll be successful.

Independent Producers

ORVILLE D. GAITHER

chairman and CEO of Gaither Petroleum Corporation

For many years, Orville D. Gaither was on a first-name basis with the rulers and energy ministers of twenty-seven countries. Now the owner of a growing independent company with two of his three sons, Gaither worked for Amoco for forty-three years. From 1979 to 1991, he served as president of Amoco's Africa and Middle East region, which in itself would have qualified as one of our largest oil companies.

The Houstonian's impressive résumé includes the 1990 presidency of the international Society of Petroleum Engineers (SPE). He holds a B.S. degree in mechanical engineering from Rice University and a master's degree in petroleum engineering from the University of Houston.

Gaither's perspective is unusual in that it combines global experience with the viewpoint of a small independent seeking to carve out a niche in an increasingly troubled market.

The comfortable Gaither Petroleum Corporation offices are located in far west Houston on a side street, somewhat hidden behind an auto dealership. Gaither seems accustomed to being in charge. One moment he is discussing the strategy of OPEC; the next he is drawing a diagram in freehand to explain how to assemble a platform in the deep ocean waters. Then he leans forward and speaks emotionally about the real image of the American oilman.

I feel that the domestic industry will be a different industry than the international sector.

The major oil companies have no real alternative except to go internationally, because we're producing only about 8 to 8.5 million barrels of oil a day in the United States. Worldwide demand is about 66.7 million—hasn't really gone up that much. During the last five years it's only gone up about 300,000 barrels a day. The U.S. is importing over 50 percent of our oil—16-16MM bpd—at the present moment.

The real fact of the matter is we ought to be using natural gas as much as we can. A lot of reasons for not using natural gas—one of them is that way back in the past the Department of Energy deemed that we couldn't deliver natural gas to companies if there were people in Boston [*sic*] that needed natural gas in their homes. Industry converted to oil and coal and, as a result of that, natural gas has suffered. But, we've got a fifty- or sixty-year supply of natural gas. We have about a nine-year supply of oil at the present moment, if we take our full demand, and obviously, we can't meet that.

For the majors, it's kind of like having a shark in a small fishpond or a swimming pool. You keep throwing minnows in there, but after a while, there are not many minnows left. Most of the majors have already gobbled up all of the small companies that they can, and they are still not able to feed their refineries. As an example, a company like Amoco has a million-barrel-a-day refinery requirement and the domestic production is about 400,000 bpd. So there's 600,000 to 700,000 barrels per day that they have to get somewhere outside their system.

Where? When I left Amoco in '91, domestic production was around 400,000 and international production about 411,000 barrels a day. Almost a fifty-fifty split of international and domestic. They must look at places that they can get into. Saudi Arabia would be a wonderful place to go, but it's tied up for Saudi Aramco and not open to the independent or major companies to do business there. We can't work in Iran or Iraq because of executive orders.

Kuwait is or has been a closed door. You can't get in there. The United Arab Emirates are one of the places you can get in, but here again it's a ninety-five-to-five split after everything. You go and figure out if you want to work for 5 percent of the gross income—that's about what it works out to be, because the split is there or eighty-five-to-fifteen plus taxes. So you end up with a ninety- or ninety-five-to-five split for the government relative to the company.

Pacific Rim

That's one of the reasons why the Pacific Rim countries are getting such a big play. The doors have been opened by Vietnam, North Korea, Thailand, Cambodia, and Indonesia. Even the Philippines have opened up again. People go where they can and capital follows opportunity. The West Coast of Africa has been relatively unstable, with all of the problems in Nigeria, the Congo, and Gabon, civil war in Liberia, and various places like that, but they are places which still seek Western investment.

I heard one of the ministers from the Congo talking at a meeting in Washington last year [1993] and—making it perfectly clear—he said, "Hey guys, don't forget us. We're out there and if our terms are not good enough for you, well, let us know and we'll change them, because we want you." And, of course, the countries like Egypt and Tunisia in the Mediterranean—I'll call them the non-sub-Sahara Africa—are pretty well opened—except for Libya, of course.

This is why most of the majors—and many of the large independents—have made tracks to the former Soviet Union. I'm not certain yet whether the fat lady has sung in the former Soviet Union or not. From 1974 to '79 I made seventeen trips to Russia, negotiating with the Russians. But when you start trying to talk to people about profit, the word is not in their dictionary. They don't know what it means. They don't have taxes. They don't have tax laws. They don't have a petroleum law. You're basically starting from scratch. And with the White Knight[1] deal, a minister comes along and puts a $5 a barrel tariff on exports and kills every cat in the alley as far as the economics goes.

These people are good oilmen, believe me. I ran into people in Siberia that, if they had been in the United States, they could have been the chairman of Texaco or Amoco or Exxon. I mean, they were first-class, high-potential oil people. But their system had them so within its grasp that they couldn't work. They couldn't do the things they knew they had to do and should have done, just because of the system.

There was no way to explain it. You had the minister of geology over here and the minister of petroleum and they would be drilling wells a half-mile apart, and neither one would know what the other was doing. Neither one could have a representative on the other one's rig floor. It was almost direct competition rather than cooperation, so the system really had them bogged down.

I think another area of promise in the international side is South America, which seems to be opening up considerably. Venezuela has some new laws that are in the talking stages, and some that are in the enacting stages. Ecuador, Colombia, and Peru have made some new rules and there is reason to believe that even Brazil may open up some of its areas. This makes it exciting for the international people, and major independents would be able to financially handle things like South America, as compared to taking pieces of things like the North Sea or China, which are much more expensive.

1. White Knight is a U.S.-based petroleum company operator that was one of the first to start drilling in the Soviet Union. It ended operations when a Soviet official abruptly put a high tariff on its exports.

Changes in skills

You're going to see a change in the type of oilman that's demanded in the world today. You're going to see a polarization, basically. I said this when I was president of the Society of Petroleum Engineers. People like ourselves—small independents—we've got to have people like my sons Doug and Duane, who can handle anything.

They handle the agreements. They handle the AMIs [areas of mutual interest]. They handle the drilling of the wells. They handle the swabbing of the wells and run the tubing. They handle the geology and they are basically generalists. In other words, they do what has to be done and they do it now. They don't submit it to a committee and fifty people go in and decide what they're going to do. There's individual decision-making at a very rapid rate and responsibility for the consequence of their actions.

Computer literacy is absolutely essential. These boys come in from the field and will sit down at their 486 and go over and start working reservoir engineering problems or creating a document for an operating agreement between ourselves and some of our partners. They have to be broad-based enough to be able to swing in a lot of different directions.

They have to be able to come in and do a reservoir engineering problem—not sophisticated reservoir engineering research—but to use the CD-ROM to pull data out of the system, evaluate that data, then decide whether to make an offer on a property or properties.

On the other hand, the degree of sophistication of the industry itself is such that the major oil companies are going to have to have specialists who are truly expert in that individual field. They're going to have to be the crème de la crème or the Ph.D. type of person that is truly a real expert in one narrow field.

I think that those areas will divide people who are coming out of school almost in the way we used to do it when we said we're going to go into research or we're going to go into operations. It's rather interesting: they either want to be an engineer and they want to be an engineer until they die, or they want to be an engineer and use it as a ladder and stepping stone to go into management.

A lot of people like myself, believe me, didn't have plans to go into management. We did our best to do the engineering job to which we were assigned, and management just came along. So we had to learn a whole new set of skills, that of being able to manage people and money as compared to being able to manage things, which most engineers are able to handle pretty well. These highly skilled specialists are the ones that would have been on the technical-professional ladder as compared to the professional managerial ladder—and I see that this new group of people that I'm talking about will be the ones who would have been in that technical professional ladder and research rather than in the mainstream of managing.

At the same time, the major oil companies are also going to need some generalists, because they're going to have to look to the person that heads up a group being more broad-based. Five or six years before I left Amoco, I initiated a team concept. We would put explorationists—namely a geologist and geophysicist—along with a reservoir engineer, and we had an exploitation team.

The last few years that I was at Amoco, these teams found more new oil than all of our explorationists put together. And the reason was that there's not much better place to look for oil than where you already are—where there's already oil. What we were doing was finding new fault blocks and new virgin reservoirs that hadn't been produced because we had a better interpretation of the subsurface of the fields.

Changes in education

One of the problems that we have—I'm on the George R. Brown School of Engineering advisory counsel at Rice University—is we see that kids coming out of high school are more poorly prepared, even though we've got a crème de la crème at Rice as far as people we get—National Merit scholars and valedictorians. But in general, we see a lowering of the SAT scores; we see people coming out of high schools that are narrow.

One of the major criticisms that's been aimed at most engineers is that they can't write and they can't make a presentation. They understand the situation very well, but trying to sell that idea to somebody else, they do not seem to do too well. So on the advisory counsel, we're looking at curricula that say maybe we ought to take another look at how we do these things.

Perhaps being able to write should be a requirement for graduation in engineering. Maybe being able to speak and to talk before a group and to be able to sell your ideas—public speaking, elocution, whatever you want to call it—to be able to sell that to somebody else, maybe that needs to be a part of the curriculum, just as physics, math, science, computer sciences, computational engineering, and biomedical. There are definitely some holes in the educational process.

Unfortunately, one of the big problems that we see in exciting people about engineering itself is that, for whatever reason, people are afraid of what they don't know. And most of them don't know basically what engineers do. As a result of that, they are not prepared because they didn't start early enough. We really need engineers and people who have an engineering bent or bent toward mechanical abilities, etcetera, be basically identified in the fourth, fifth, sixth grades, not by the time they get to high school and college, because by then they haven't taken the physics and the math and the courses they need to be accepted into a first-class engineering school.

Waking up to 3-D

Three-D seismic is uncovering tremendous amounts of new production within old fields and fault blocks that were missed, pays that were bypassed, pays that weren't even known that were below the depth of the field sands. These things are revitalizing particularly the domestic industry, but they have also revitalized the international industry. We've found new fields in Egypt—Amoco has within the last few months—basically an old field but new fault blocks and new segments that were not known before the advent of running a 1,200-mile geophysical survey in 3-D.

I think that you're going to see an expansion of the 3-D into maybe the fourth dimension, where the fourth dimension would be reservoir analysis. For example, in the days of yore we used to have what we called the reservoir limit test. In it, we used pressure transient analysis, namely the draw down, and you saw pressure behavior to predict the presence or the absence of a fault or of a barrier and to basically give you a limit to the areal extent of the reservoir.

They're now working with this type of thing in conjunction with 3-D seismic to predict not only where the faults are, but the size of the reservoir. This is a whole new field. Here again, it won't happen except through team efforts, because the geophysicist is going to be able to understand 3-D, but he's not going to understand pressure pulse testing and that sort of thing. That's why I see a lot more relationships between the interdisciplinary groups working in petroleum engineering.

Before I left Amoco, I had told them that there were two things that needed to be done to be a preeminent oil company. One was to be able to image below the salt geophysically. I guess you've seen some of the more recent data where Amoco, Anadarko, and Phillips have come up with interpretation methods to be able to get energy through the salt. Usually salt is kind of like a big old sponge—it absorbs most of the energy and it's hard to get energy through salt. But, having said that, being able to image below the salt is something that the industry is going to do. It will be one of the major breakthroughs to the industry for all times.

The other thing that you basically need to get a leg up on is to be able to figure out ways to recover more oil from existing fields—CO_2 injection, for example. We talk about doubling the oil recovery, but maybe we're talking about doubling a 17 percent number to 34 percent, or we may be doubling a 30 percent number to 60 percent.

The fact remains that we've probably got more oil left in the ground today than we've found, simply because we don't know how to get it out economically. Whatever you replace that oil with has to be cheaper than the oil itself, and at $12, $13, $14, $15 a barrel, there's not many things that you can buy for that price.

A Saudi "victory"

I think that you'll see a continued dominance in the oil industry of the OPEC countries, because they do have 66 percent of the world's known oil reserves. These foreign suppliers are still controlling the industry and will be for a long time to come.

Talking about this book, *Victory*,[2] a few years back, I was in Saudi Arabia as the guest of Saudi Aramco. A friend of mine, Dr. Sadad Husseni, is executive vice president of Saudi Aramco. He is a Ph.D. graduate of Brown University and former board member of the Society of Petroleum Engineers. He was telling me that the Saudis were going to increase their producing capacity to 10.5 million barrels a day, and it was going to cost them $2.2 billion to do it.

I said, "Sadad, I just don't understand. Why in the hell would you want to do that when you're only producing 5.3 million and you've got a capacity of 6.5 or 6.6 million barrels a day?" He said, "We've got to be able to do it in case any of the other OPEC members falter in their ability to produce. And, besides, the king wants to do that."

That latter statement didn't make any sense until I read *Victory*, and now it makes pretty good sense. If there were some sort of a deal struck back then, the king would have been privy to it and nobody else in Saudi Aramco would have been. So he just had issued an edict to increase the production capacity to 10.5 million barrels a day. It's that simple.

We're going to be, in some respects, dependent upon those people for the pricing. It costs Saudi Arabia about 94 cents to produce a barrel of oil. They've got so much oil already found, yet they're out with multimillion-dollar contracts to go back and do 3-D seismic over a lot of their old fields and their new fields and the area particularly in the Empty Quarter. But there is no doubt that they've got tremendous undeveloped reserves.

There are not any places in the world where we can operate that cheaply. The North Sea production is where you've got to strain to produce oil for $15 a barrel to make any money. Russia is a place where potentially you could be able to produce oil at a very low cost because of the huge volumes involved. China, I don't know; I've kind of stayed out

2. Peter Schweizer, *Victory* (New York: Atlantic Monthly Press, 1994).

of China. But it's hard for me to believe that the Chinese, as shrewd as they are, will ever let you make any real money in China.

At the same time, technology is on our side in that we've had some tremendous breakthroughs. One of the things that bothers me is that so many companies are curtailing research activities. This may force research back into the universities. Maybe that's where it ought to be. But most of the private universities would find it very difficult to fund it. State universities are not usually known for being hotbeds of research. I see that as being a change.

The oil industry still is one of the most exciting businesses in the world. There's so many different facets to it. Every time you think you know all there is to know about it, it's like a light hitting a diamond a different way and you see a little glint that you have not seen before.

Casting in deep water

It's going to be tough to make a living in the future. And it's going to be very capital intensive, maybe more so than it has in the past. Deepwater exploration—you talk about the difference between a 100-foot platform or a 500-foot platform and a 1,000-foot platform, and then you start talking about producing in 3,700 or 4,000 feet of water in a typhoon belt, and costs escalate exponentially.

Amoco's South China Sea project is an example. You would be hard-pressed to find a more difficult place to work: in a typhoon belt, in 1,000 feet of water, the reservoir is 1,800 feet below the surface or 800 below that—not below the surface of the water, but 800 feet below the sea bottom—with a very highly porous limestone that has the ability to transmit the active water that underlies the oil field almost without any resistance whatsoever to vertical flow. It's just infinite permeability. You must drill horizontally, and there is insufficient reservoir energy to be able to produce the oil without electric submersible pumps. These pumps are designed normally not to be laid on their side, but to be hung—they've got thrust bearings but not radial bearings. So now the pumps have to be redesigned to have thrust bearings and radial bearings as well in order to lay them on their sides.

And then couple all this with being in the typhoon belt, where you've got to be able to get loose from the producing well and to go back and reconnect to it, and you're talking about astronomical sums of money. I was reading about a local company the other day. Two major finds—11,000 and 12,000 bpd wells—and the stock jumped up and traded six times the volume at the same time and the price stayed exactly the same. Why? I strongly suspect that people would look and say, "Does a company like that have the capabilities monetarily to be able to develop a find at 3,000 feet of water?" You can drill it, but there's a lot of difference between drilling it and putting it on production.

No alternatives

If you put it purely and simply, we don't have any alternatives to fossil fuels at this moment. Maybe a hydrogen economy or something else with a fast breeder reactor could happen. Already countries like France generate 85 percent of their power needs from nuclear sources. This country is scared to death of nuclear, and yet we've never had an accident that's killed anybody in nuclear energy.

Maybe we shouldn't have stopped the development of the fast breeder reactor under the Carter administration, but Jimmy being a nuclear physicist, I guess, knew some of the pitfalls that maybe the rest of us don't know. But you look around, you say, "Well, we will make ethanol and mix that up with gasoline." You could do that to an extent of 20 to 30 percent, but ethanol is hydroscopic, and materials that we make our automobile engine out of will still rust. When you start putting water in your system it could cause some severe problems. We don't know what the emissions from a mixture of ethanol and gasoline are going to be—maybe more environmentally unacceptable than reformulated gasoline. We don't know the answers to that yet, and may not for years.

Can we be energy self-sufficient from fossil energies? To me, the answer is "no." You've got two-hundred-plus years of energy capabilities as far as coal goes, but I just don't see coal being a very mobile fuel. People talk about an electric car—these same people, you ask them where electricity comes from and they say, "Out of the plug on the wall." They don't have a clue that it's generated somewhere, usually by fossil fuel, and has to be transmitted through wires and into that plug. They just feel like it grows on trees, I guess.

But if you have to generate electricity from fossil fuels, which is about the only alternative unless you want to go to nuclear, that efficiency is about 35 percent, and it's even less than that. You put that into a lead acid battery that can be recharged a number of times, but it has a shelf life of maybe one to two years. At the end of that time, it's got to be disposed of. So you've got the cost of remediation of the lead acid materials, all of which is now hazardous waste.

Sanity needed

The only other real major concern that I have is, I think in this country the oil industry will ultimately be destroyed by the environmental program. You can't be environmentally insensitive, but you can be environmentally sane. For example, the concern over recycling of plastics. You say, "Plastics are no good. They clog up the landfills and all these other things." And yet, that's true, but if you really believe that the greenhouse gasses are more dangerous than something else, it [recycling] uses more energy and costs more from an energy standpoint—burning fossil fuels to make plastic—so it costs more to recycle it than it does to make it.

Some of the rules are absolutely unbelievable. They are mandated by people who are not scientists and have no real basis in fact for their action. We've just bought a field out at South Texas and are going through the remediation process. Over the years, mercury has been blown out of meters and maybe an area underneath the meters that might cover eight feet by eight feet, or sixty-four square feet—there may be mercury contamination.

The question is, does it make it better if you dig that mercury-contaminated soil up and put it inside a sterile plastic barrel, and ship it to New Mexico and let it sit there in a concentrated area, or is it worse to just let it be there? First off, it's going to ultimately go back into the ground where it came from. So, what do you gain by moving that away? It's the rule—it's a hazardous substance.

[Son Duane Gaither has walked in and taken a seat.]

Duane: We believe in environmentalism, but not rabid environmentalism. A case in point is that last year, a company that we have been involved with had a hole in a tank. It was a gun barrel used for storing salt water and oil. And the oil squirted out of the gun barrel—the oil and salt water squirted out of the gun barrel across the dike that was built to contain such a catastrophe, and ran down into Spring Creek. The Texas Railroad Commission began the oil spill mediation process and eventually traced the oil back to this operator.

When all things were said and done, they spent $56,000 recovering approximately seven barrels of biodegradable crude oil out of Spring Creek. The pads and the absorbent sponges and booms that they used to gather that oil were out of a man-made material quite similar to a diaper. Once oil adheres to that, it closes up like a wet diaper and that material becomes toxic waste. So we took seven barrels of biodegradable crude oil and converted that into seven tons of nonbiodegradable toxic waste, which is enough to more than fill this room.

We didn't fix the problem. We compounded it by over tenfold by being irresponsible. Now this is our government doing that. Our response has been to not fund the more rabid groups, but to fund the groups such as ARE [Association for Responsible Environmentalists].

Orville: I think the goal of some environmental groups is to absolutely have no production in the United States—no production onshore or offshore. They've already closed the East Coast. They've closed most of the West Coast. They've closed most of Alaska. They've closed all of the Gulf of Mexico back around to Panama City, Florida. They're working on the western Gulf at the moment. Their goal is to shut down all operations offshore.

Danger of spills

It seems like a bleak picture, except that if the good news is we made some kind of a deal with the Saudis to keep the price down, maybe the American people will continue to get a cheap thrill out of energy. Certainly, they're not paying their own way now.

You hear people talk about an import fee and you hear people talking about the cost of double-hull tankers. There's no doubt about it: the more we import the more chance we have for major tanker spills and a repeat of the *Valdez*. That's very significant, because as we grow more and more dependent upon imports, we're going to see more and more tankers in U.S. waters.

Basically what we have is a policy not to have a policy, and maybe that's a policy in itself. You understand what I'm saying? The United States has said we're not going to have an energy policy. We haven't built a refinery in twenty-six, twenty-seven years, and we probably will never build another one because you can't get the environmental permits to do so. Our policy seems to be import and phase out domestic production.

I feel we're going to see more and more imports of finished products like gasoline and diesel fuel coming out of places where there are refineries, such as in the Caribbean or Saudi Arabia and Kuwait, where they'll be refining and shipping products to us in a more volatile form and it will cost more.

With $10 or $12 oil, there's not a lot that you can do to have a viable industry. Let's say, for example, we've determined that the industry can be viable in the United States at $25. We'd have to put an import tariff of $10 a barrel. Who would the $10 go to? Would the $10 go to the federal government? If it is, it's not going to help the industry at all—except that it would raise the price in the United States to $25 a barrel. Now, if you put it at $25 a barrel, people are not going to be able to buy gasoline at $1 a gallon. It's going to cost $1.40 or $1.50.

Cheap oil at all costs

The people of America don't want high-cost energy. It isn't a politically viable thing. You'd have a terrifically difficult time getting it through the Congress.

The oil industry has lost 500,000 or 600,000 jobs, and there's very little hue and cry about it. The coal industry lost about 200,000, and the automotive industry has lost about 300,000. And both of those are highly vocal about moving our jobs overseas, you've got to restrict imports, cars, and so forth. I guess if I had to make a choice, I'd rather see less regulation and more freedom to be able to do the things we do best.

Look at the airline industry. They screamed and hollered for deregulation, and then once they got deregulation they didn't want it. That may be the case for our oil industry. Maybe we want to be deregulated, but we may not be as happy without it as we are with it. I don't know the answer to that.

Energy is unique in that it is not something that happens instantaneously. It's an industry that takes a long time to build. You can build a Compaq computer in a matter of minutes, and you can build a company in a matter of years. The oil industry, you build it over a factor of tens of years and twenties of years, maybe hundreds of years. And its destruction can't be replaced. Once a stripper well is plugged, you can never afford to go back and redrill it.

Some of the things that the Texas Railroad Commission is doing, saying, "Don't abandon these wells, we'll give you an extension on keeping them in place, we'll reduce the severance tax on stripper wells under three barrels a day," and things like that, will help preserve the industry for a little while. But just preserving what we have is not enough to keep the industry alive.

Those things are helpful because, who knows, ten years from now we may have found out a method that we can recover all of that oil if we just had the wellbore, but if we've abandoned the wellbores, we will not have them and we cannot afford to go redrill wells because of the economics. Some of the things that are happening make very good sense in that respect.

If you look at it from a national defense standpoint, such as your ability to wage a war, you might get a totally different picture. It seems the Pentagon has decided that you're not going to see global wars, you're going to see brushfire wars. Maybe energy is more worldwide than it is localized. So, as long as you don't have a madman like Saddam Hussein in charge of Kuwait and Iran and Iraq and Saudi Arabia, well, maybe we do not have a problem.

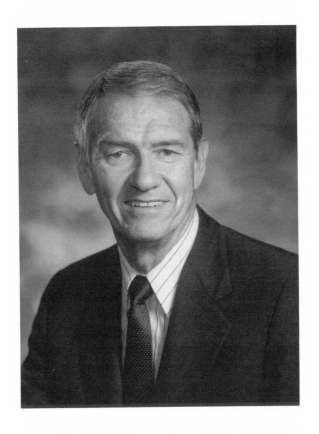

JOE B. FOSTER

chairman and CEO of Newfield Exploration Company

Joe B. Foster is founder, chairman, and CEO of Newfield Exploration Company, a recent success story in the industry.

For thirty-one years Foster worked for Tenneco Inc., Houston's biggest home-based corporation. He became chairman of Tenneco's small but highly regarded oil company, which had operations spread throughout the world. But in May 1988, the board of directors of Tenneco Inc.—with Foster vigorously dissenting—voted to sell the oil company and use the proceeds to salvage some of its less-profitable companies.

The oil company was put up for bid in September, delighting competitors, which eagerly gobbled up its assets. Foster, given the opportunity to bid on the company by Tenneco's board, arranged financing, and made unsuccessful offers for three of Tenneco's packages. Determined to run his own company, he began looking for initial investors. In late November 1988, Newfield came into existence.

Today, Newfield specializes in exploring for and acquiring oil and gas reserves in the central Gulf of Mexico. In 1994 and 1995, it was included in *Fortune* magazine's list of the nation's "100 Fastest Growing Companies." In July 1995, Newfield was also listed at the top of *Texas Business* magazine's list of "100 Fastest Growing Texas Companies." In 1994, Foster was named "Entrepreneur of the Year" in Houston's energy sector as part of *Inc.* magazine's annual competition.

A native of Texas, Foster holds degrees in petroleum engineering and business administration from Texas A&M University. He is highly respected throughout the industry, and it is easy to see why. While keenly interested in discussing his industry, Foster also shows a genuine concern for others. He is committed to learning and exploring, whether in the depths of the Gulf of Mexico or on the highest mountaintops of Africa.

*T*o me, the key to an ongoing, viable industry is letting the market work. This has not been a market-driven industry. Historically, we've had price managers: the Texas Railroad Commission, the Oil Import Commission, OPEC. I was born in 1934, and from 1934 to, I believe, 1980 or '81, the nominal price of oil never went down, so you've got a whole generation of management that really never worried about price risk until the 1980s came along. On the natural gas side, we had gas that was sold at regulated prices.

Now we're learning to deal with the price risk, along with all the other risks that go along with our business. As we've seen deregulation and more market forces take over, it's been a tough transition for this industry. In the end, however, I think that's going to result in the necessary supplies coming forward, and to the extent we can move this industry more in that direction, that's probably the thing that needs to be done most.

Role of government

That's not to say government doesn't have a role. There will always be safety and environmental issues where regulations are going to be required. There will be level playing field issues that the government will have to address. But I'm not sure that the government needs to subsidize this industry to accomplish what it needs in the way of adequate supplies.

Probably less regulation and more enlightened regulation is something that the industry needs. Both the government and the industry itself have moved in the right directions in the past few years. That's not to say that I like all the regulations we've had. There have been some unduly harsh environmental restrictions placed on us, and that needs to be addressed.

The move toward the deregulation of natural gas prices, the move toward all the unbundling that has taken place as a result of the Order 636, have been good. I think the whole gas industry is a much more competitive industry, with room for more players and more people to take advantage of competence, than it was previously. Despite the fact that a lot of producers feel gas prices might have been higher had they remained regulated, I think the industry is much more vital.

On the oil side, going back to Reagan's administration, it did get rid of entitlements and other kinds of things that created unnecessary bureaucracies and more disincentives than incentives. Again, the move toward the market has been good.

What's not been good are such things as the Oil Pollution Act of 1990, which came in the heat of the *Valdez* and was very much an overreaction. If OPA '90 is enforced the way the legislation is written, it would require a company like Newfield to have a $150 million Certificate of Financial Responsibility, which we couldn't get, in order to operate in the Gulf of Mexico. Hopefully, that act will be modified. I think the present Congress has some intention of doing that and companies like ours can stay in business in the Gulf.

Regulators have another tendency that annoys me. Every time there's some new technology that comes along which lets you lower the number of parts per million of this or that, they lower the standard, even though it's not necessary. They're not really saving anybody's life or improving anybody's health. But because it's technically possible to do, they require us to do it. There is not enough attention to what they in Washington call cost-benefit analysis in many of the regulations that are being promulgated.

Government subsidies for alternative fuels are, in my judgment, unnecessary. Let there be interfuel competition as well as industry competition, within a reasonable set of tax laws. Let the market figure out which is the better kind of fuel as opposed to trying to subsidize one or another. The market works!

I recognize the arguments about the national security issues, and they really ought to be considered, to the extent we incur a lot of expenditures for defense or for whatever reasons in order to assure access to imported oil. We ought to take that into consideration as to how we deal with domestic oil. We need also to consider balance of payment issues.

I don't basically like the notion of tariffs. On the other hand, if we have a demonstrable cost that we incur because we are importing oil, and some of those costs could otherwise be incurred to bring forth an equivalent amount of domestic oil production, then we ought to look at mechanisms for doing so. It would be better to leave those dollars in the U.S. than to send them abroad.

I don't know whether that sort of rigorous analysis has ever been made. The import solution is usually presented as "we need to subsidize domestic oil." It ought not to be a subsidy. It ought to be an economic trade-off between domestic oil and foreign oil at the macroeconomic level.

Forward markets

The move toward the creation of financial instruments to let us deal with price risk has been good. This gives us a way to deal with the fact that we have to invest so much capital up front, and then once we commence production, our operating cash margins are relatively high. If we get in a battle where incremental costs are what's determining price, it's very easy to reach a point where we are selling our product at lower than the all-in cost, including capital cost. That's why the Texas Railroad Commission and others have tried to regulate volumes over time: operators were creating excess supplies with over-investment and reached the point where they were selling their product at an all-in loss, but at a current cash profit.

One way to address that is regulation—proration or cartel. Another way to address it is to have financial instruments that let us sell our product forward over a longer time frame, and, in effect, manage that price risk that way. Those avenues are now much more generally available than they were even five years ago.

A company like Newfield, for example, which has no marketing department, no transportation arrangers, can hedge or make swaps with respect to our production so that we can manage the price risk fairly effectively. That's a very healthy ingredient to go along with deregulation of markets.

Changing hands

There has been a lot of positive changes as a result of the restructuring that has taken place, and the industry needs to continue that process. It is important for there to continue to be the kind of recycling of properties that Newfield has participated in, where marginal and lower-tier properties which are no longer of much priority or profitability to the bigger companies are sold or farmed out to companies with different kinds of cost structures and economic incentives and are able to find additional oil and gas. What is healthy is for these properties to change hands.

Consider what happens with a lot of those acquisitions—it has certainly been the case at Newfield—that a lot of exploration takes place on those acquired properties. I was asked by a security analyst to compare the profitability from pure exploration versus the profitability from acquisitions. I said I really can't divorce the two, because a lot of good exploration prospects are on producing properties. That's the major reason we acquire producing properties—to gain access to prospects that we could not otherwise access.

There has been a good relationship between the majors and independents through the years. Majors have used the independents to get a lot of wells drilled. I think the statistics will probably show that the independents drill more wildcats in the U.S. than the majors do. In the past, however, there have been any number of times where the larger companies have held on to property that could have been better exploited by a smaller company, property that just sat there producing less than it could, being less explored than it could, because it didn't have a very high priority in the scheme of things with the majors. Many of those properties have been offered for sale by majors and bought by independents since 1986, when oil prices plunged, and the restructuring and downsizing movement began.

I had a major company executive tell me a few years ago about a property they offered for sale, but didn't receive bids which were as high as the present value of continuing to produce the property. Upon asking, he found their net present value was based on them drilling three wells within the next year. He told his people, "These are wells that have been on our budget for ten years and we have deferred drilling them every year. Run the economics on the basis that we don't drill the wells for five years. Then what's the net present value?"

It turned out this net present value was lower than the bids they had received. That really is a proper way to look at it. Once you decide that "Yes, there's a drilling opportunity there, but I'm not going to get around to drilling it for five years," you maximize present value for your shareholder by selling the property rather than keeping it. There needs to be more of that kind of discipline in the companies that have a lot of properties.

In the grand scheme of things, many of these are opportunities that don't mean that much to a major, but they mean a great deal to a smaller company.

I think some independents will be consolidating, yet there'll continue to be new ones formed. It's sort of a bubbling process. There'll be new companies formed that have specific competencies and skills, like Newfield. At the same time, there'll be other companies that have outlived their usefulness or, for one reason or another, want to cash in their chips, that will get merged or sold into other companies.

There are opportunities overseas. The same things work over there as work here, except it takes more political skills and it takes more patience, and probably more capital. Independents are normally not long on any of those: patience or political skills or capital. But as the world gets more and more compact, transportation wise and information wise, you'll see more independents overseas.

Natural gas will continue to play a major role for independents, mainly because that's generally what's left here. That's not to say there's not going to be quite a bit of oil yet to be found. But since so much gas was bypassed in the past because it was priced so low, industry will go back and find some of those deposits. As we drill deeper, we're more likely to find gas than oil, just because the temperature of the Earth is greater at depth, which would lead to gas being there instead of oil.

Oil is still the basis for transportation use. We are, for a very long time to come, going to have a very substantial part of our energy being supplied by oil. I cannot see electric cars taking over. I'm not a strong advocate of CNG—compressed natural gas—for transportation. There are a lot of infrastructure problems, in addition to the inertia of the bigger companies that have established gasoline marketing outlets. They're not going to be real cooperative.

Lost image

The industry is a hell of a lot more responsible than it's given credit for in the press and by the public at large. I don't know anybody in this business that really is so interested in making a buck that they're going to deliberately foul the air or the water. That's just not the way it works.

My theory is that when service station attendants quit wiping your windshield, that's when the American public lost confidence in the oil industry. I remember telling several people in my hometown when I was getting out of college that I was going to work for this company called Tennessee Gas that nobody ever heard of. I had several say, "You really ought to go to work for Humble. Old so-and-so has been taking care of my car for years, and they really do a good job. Humble is a terrific company."

It depended on who that service station operator was as to whether they thought ABC or XYZ was a good company. Now, it really is a "filling" station, just a faceless place. There's not even an attendant. You just stick your credit card in the pump and get your gas. I really think the loss of personal contact at the service station has been a big part of it. ·

Success through technology

Most of our success has come from the application of technology. One of the things that we set out to do when we founded Newfield was to apply major company technology in the Gulf of Mexico, coupled with an independent's cost structure and mind-set. We started out with $9 million of equity capital. Two million dollars of that was spent on Day One buying geophysical data for the central Gulf of Mexico. We spent another $200,000 or $300,000 buying a Landmark workstation so we could work 3-D geophysics. We spent another $150,000 buying computers and software so that our engineers, geologists, and geophysicists would have the tools they needed to manipulate this data.

Although that sounds like a lot of money, we built that technical base at a lot less cost than I'd experienced getting the same level of technology with the bigger company I worked for previously.

We have been helped greatly with the advent of 3-D technology. It has let us see some things on properties we simply couldn't see previously. We've been helped with the advent of horizontal drilling. That's let us achieve productivity that, again, wasn't previously possible. The real trick in this business is to use a very high level of technology, to use it rapidly to establish production and revenue quickly at high rates, and then to keep costs extremely low. If you couple low-cost structure with high-technology utilization, then you've got something.

It's possible the next big technological advance could come in formation evaluation, where we might be able to see some things in old wells and old fields that we hadn't seen before. Probably the greatest technical advance that could take place there would be improved recovery efficiency of oil. This got a lot of press during the high oil price days and the energy crisis.

There were a lot of incentives directed at secondary and tertiary recovery. However, not as much progress resulted as the dollars spent should have produced. Still, there are billions of barrels of oil that have been left in the ground. A way to get even some of that out could add immensely to our country's reserves.

Newfield's game plan

In our original business plan, we wanted to give ourselves the benefit of every advantage we could. We felt that focusing our efforts and attention in the area with the greatest hydrocarbon potential in the United States was the way to give ourselves the best chance of succeeding as a start-up company. We've done very well in the Gulf of Mexico.

We could acquire technology much less expensively in the Gulf of Mexico as a start-up than we could onshore, because geophysical data is less expensive to acquire in water. There was a lot of speculative 3-D as well as 2-D data, so we could purchase data that had already been shot and the cost would be shared with other people.

Tenneco had properties all over the world. So we did know a lot of the areas. I wanted to start in the Gulf of Mexico because I thought there was more potential, the barriers to entry weren't that great at that time, and the technology was least expensive there.

Tomorrow's leaders

The major oil companies and the American Petroleum Institute have always had a very strong voice in this industry, for good or ill, and I expect that to continue. They have a tremendous amount of capital tied up in this industry. Other trade associations will have important voices in that leadership as well. The IPAA is a good organization that helps communication take place in the industry and gets its views known.

I don't look for a Lee Iaccoca to emerge in this industry, because it's such a highly fragmented industry. The best we can ever hope for is some coalitions where specific organizations work together on specific issues.

If you talk about leadership in the way the industry is managed, it will be by example more than anything else. One of the issues that exists in this business is that oil companies have become too bureaucratic, and I've thought for some time, particularly since I've been at Newfield, that the central managements of the bigger companies need more of an investor orientation than a controller orientation.

We have a large venture capital firm which has a significant investment in Newfield. It has two directors on our board. They've had a great deal of influence in helping us shape our strategy and thinking about how you run a start-up company and what sort of opportunities we should pursue. But they are strictly investors. Their suggestions and ideas are directed at helping Newfield make a good return. That's what we're in it for, too, and it's worked well.

This is in contrast to a situation where some people from corporate staff come in, set our budget, have the power to withhold funds, and are continually issuing report cards to us or about us—or, are asking for reams of data from us. That's one of the reasons smaller companies can be successful in areas where bigger ones may not. There's an investor means of control as opposed to a bureaucratic means of control.

The industry at large sees some of these benefits, and I think when people talk about "autonomous business units" in these bigger corporations, that's an effort to move in that direction and that's a positive effort. Look at what Enron has done. They've got a lot of subsidiaries that are partially owned by the public. So you've got investor attitudes that go along with the overlay of Enron's parent company ownership. That's a way to better manage large organizations. It may or may not be the best model, but it's one that other people will at least look at. That's a form of leadership that is not anything other than just being a little bit out in front in the way they do business.

In a very small way, Newfield has been a leader in causing other companies to reexamine the potential of the Gulf of Mexico and the way they operate there. I think there will be that kind of leadership in this business, where corporations simply look to other corporations and learn from them. That will be an ongoing process.

GLENN H. MCCARTHY JR.

chairman of McCarthy Oil & Gas Corporation

Glenn H. McCarthy Jr. is a third-generation oilman whose company, McCarthy Oil & Gas Corporation, carries the name of one of the most famous individuals in the history of the petroleum industry. Both of McCarthy's grandfathers worked at Spindletop when it was brought in shortly after the turn of the century, and the McCarthy name is firmly entrenched in the roots of two of the great oil companies, Texaco and Amoco.

McCarthy's flamboyant father, who died in the late 1980s, was a legend in his own time. Glenn McCarthy Sr. was regarded as one of the world's greatest wildcatters, even though he wasn't an engineer or a geologist. He is the subject of a book now being written by his son, who describes him as "probably the greatest natural oil and gas finder that ever lived."

A soft-spoken, burly man in his mid fifties, Glenn McCarthy Jr. is also a lawyer, having graduated from the South Texas College of Law. He was extremely well prepared for our meeting. Seated behind the family's partner desk that helped shape the history of the second boom at Spindletop, and beneath a framed copy of the *Time* magazine issue that featured his father, McCarthy spoke about the concerns that trouble this country's dwindling number of independent producers.

The lack of price stability concerns me. With the current state of price variation on oil and gas, it's impossible to forecast with any degree of accuracy the rate of return on investments, and to formulate future drilling budgets, for lack of certainty on revenues. When you're proposing a venture with someone to drill a well, obviously they want to know, "When am I going to get my money back?" They want to know the risk factors, among other things. We can estimate volumes of production with a great degree of accuracy, but while prices fluctuate greatly, the revenues will fluctuate accordingly.

Personally, I never have been an advocate of any import tax on foreign oil. I don't think that's the way to solve our problems. The biggest problems are the governmental regulations that stymie us everywhere we turn; the punitive aspects of the taxes that have been placed on the industry; and the reduction of intangible drilling cost deductions and passive and/or active income classifications, which have extremely detrimental effects on fund-raising efforts. If it's passive income, then you can't use losses generated in an oil venture to offset that. That directly precludes a number of investors in the industry. Attorneys or doctors—some have very large incomes, but if they don't have direct oil incomes, they can't use their losses against their passive income, as it is classified.

It's very difficult to talk to someone when they say, "What if we drill a dry hole?"—which is always a possibility whenever you poke a hole in the ground. "Am I going to be able to write it off?" All you can logically say is, they've got to talk to their own tax people. None of us want them relying on our interpretation of taxes for liability or practical purposes. The fact is, we don't know their intimate tax situation and we don't want to know it.

Alternative minimum tax

But what you've got to say if you're going to be truthful with them is, "Yes, if it's passive income, you're not going to be able to write it off." Then, what about the alternative minimum tax? The alternative minimum tax, in effect, if you want to take it to its nth degree, means you can end up paying taxes on a dry hole.

Theoretically, if you had a million dollars income and you drilled a million dollars worth of dry holes and you had zero income that year, mathematically that sounds right. But according to the alternative minimum tax, you have to apply the alternative minimum tax against the income. If it's 15 percent now as the alternative minimum, you still have the situation where you have a million dollars income and a million dollars of losses, and zero income, but you still have to pay 15 percent on the income, which means you owe $150,000 in a year that you had zero income.

It has been ameliorated downward somewhat; so if you reduce it, that makes it fair? I don't care what you reduce it to: if it's 28 percent or 15 percent or 1 percent, it's 100 percent unfair. It's selective to the oil industry and is not only nonsensical, but punitive in its nature.

Again, if you talk to a potential investor in an oil venture and they say, "Is that the truth? Can that happen?" You have to say, honestly, "Yes, it can." You would damn near have to be brain dead as an investor to go into a proposition like that. It makes it almost self-defeating to try to talk to someone about drilling a well, because of the governmental regulations and taxes.

All in all, the oil industry is the heaviest taxed industry in the United States, probably the world. We pay more taxes than any industry there is and traditionally have. We pay severance taxes, ad valorem taxes, income taxes, transportation taxes, storage taxes. We've got alternative minimum taxes. It's almost impossible.

And what about the fluctuation in prices? We've all heard many different proposals about a floor, that if oil goes below a certain price per barrel, then they'll add a tax to it that will bring it back up to whatever floor price there is. In my own mind, I have not determined what I think would be the best way to do that. But there's got to be some way between industry and government that we could come up with some sort of palatable solution that they can somehow tell us what the price of oil is going to be.

If it was $15 a barrel, that's fine, but we need to know. I would a lot rather it be $15 a barrel now than $20 for two months and $12 for another two months. It presents an impossible economic situation to try to survive in that environment. The government protects many industries—auto, timber, fishing, all types of agriculture—but seems unconcerned about the oil and gas industry.

Reagan's "victory"

There's a book that was recently published called *Victory* that I thought was very interesting.[1] It set out the orchestrated attack by the Reagan administration on the Soviet Union, which they called economic warfare, which it was. Their purpose was to cause a collapse of the Soviet Union, which they did. If you've read the book, you're familiar with what I'm talking about on several different fronts. The covert aid to Afghanistan and the Solidarity Union in Poland to increase the cost to the Soviet Union to combat those areas. The misinformation that was given them because of their stealing secrets from us, so we gave them bogus secrets, which caused their pipelines to be built incorrectly, costing huge amounts of money.

But the whole basis of their [the Reagan administration's] attack—the economic warfare, from reading the book—was the deal they made

---

1. *Victory: The Reagan Administration's Secret Strategy That Hastened the Collapse of the Soviet Union,* by Peter Schweizer, (New York: Atlantic Monthly Press, 1994), purports that the administration made a secret deal with Saudi Arabia to flood the market with cheap oil, this depriving the Soviets of desperately needed export revenue. At the same time, the administration rejected Soviet efforts to acquire U.S. technology that would have alleviated problems in its energy sector's deteriorating infrastructure.

with the Saudis to keep the price of oil down in return for military protection. That was to the direct damage and detriment of the domestic oil industry. In the interviews in the book, different Cabinet members and major players readily admit that. They admit they knew at the time it was going to cause severe damage to the industry.

Most of them, to my reading, said they did not realize the ripple effects of the damages to the industry. They didn't realize it was going to fall over into real estate—cause a collapse of real estate—and into banking, leading to the S&L banking scandals. But in my reading, I felt they were saying that, even so, we still would have made the same decisions because of the goal that we were trying to achieve, which was the destruction of the "evil empire."

In my position in this industry, what they did doesn't sanctify it. It does, in a sense, make it more palatable. That if we were casualties of war, so to speak, at least the goal was accomplished. The Soviet Union no longer exists. That doesn't help us economically. The damage is still done. But at least to know the damage that was done to us did accomplish the goal that they set out.

I'm sure most of us in the industry knew that what was happening to us was orchestrated, although we didn't have the whole picture put together as set out in that book. But I think we all knew it couldn't have been a mistake—it would have had to be orchestrated—the downfall of the industry.

Government regulations

As far as government regulations, it seems as though every time we turn around, there's another regulation, another restriction upon us. Ecologically, it's gotten to a point where it's like a bunch of rabid animals that have come to a panic stage. And the things they do are nonsensical, counterproductive, and don't help the environment a bit. Taken to the farthest degree, if a migratory waterfowl is found dead in your mud pit, it's a $25,000 fine. So we all have to put nets over our mud pits now.

That's good and bad. There are other things that get under the nets and might die, too, and who's to say that a dead waterfowl was not wounded by a hunter and got in there and died? But they don't care. That's what I'm saying. They're so hard and fast and the rules are so black and white, it doesn't give you any leeway. And that's just one absurdity. I once drilled a well in the Sam Houston National Forest. It was an area where the landowner had donated the surface of the land but kept the minerals. The federal government then owned the land. I had to get seventeen separate permits to drill that well. I had to spend tens of thousands of dollars on obtaining the permits.

It was in an area that was an old oil field. There were old pump jacks all over it. There were exposed pipelines that were rusted and old. It was drilled many years ago and abandoned many years ago. The storage tanks are still there. But they had classified this as a Rare II area: roadless areas of review and evaluation, which are supposed to be just backpack. That's how foolish they are. It wasn't like going into a pristine forest. It was actually in an old oil field. Yet I still had to comply as though I was going into a pristine area.

In that same instance, we were told by the Audubon Society that we had to move our access road to the location. In the road to the location was an old dead oak tree which they said had a colony of red cockaded woodpeckers. I asked them at the time, "How many is a colony?" They said, "Two." We had to move our road, I believe it was 1,600 feet from the tree—go around it. Because of doing that, we had to go through two creeks, build two bridges—which added $50,000 to the cost of the location.

Maybe the ecologists say, "Who cares? That's your money." Well, sooner or later we've got to get that money back, so we have to charge higher prices, which causes the public—John Q. Citizen—to pay more money. It costs all of us money every time they do something like that.

I'm a hunter, an outdoorsman. I love the outdoors, nature. I care more about the preservation of animals and wildlife than the ecologists do, because I want my children to be able to see them.

And, incidentally, hunters pay the cost for the environmentalists by fees and taxes on ammunition and so forth. We're much greater ecologists and conservationists than they are—the ones who claim to be. But that movement cost us, in other words, $25,000 per woodpecker. Personally, I don't give a damn whether those woodpeckers are alive or not. It infuriated me to have to do that.

Again, that's one small incident that we have to contend with. In order to do that, it takes time. If they don't want you to proceed, then they put on the brakes. They do everything, not in slow motion, but in reverse. To get one forest ranger to sign the document that we had the right to go in there after all these permits—if he goes out on vacation, we sit there and wait. And it's not because he was scheduled to go on vacation, it's because he didn't want you to do this. It's the attitude. It's not a matter of lack of cooperation; it's a positive movement against you to keep you from doing things that you are legally entitled to.

That's just a couple of things that have happened to me. You could sit down and document hundreds of instances like that, that cost millions and millions of dollars, and it's not productive. It's counterproductive. Especially, like I said, this whole thing was drilled in an oil field. And they treated it as though it was in Yellowstone National Park.

Another thing: we've got to have some sort of parity, equality, where we're treated like normal human beings. We're treated by ecologists as though we're criminals. That's the way they treat the industry, and that's the way they look at us. The men and women that I've known in the oil industry are dedicated, conscientious, marvelous, educated people who strive to do their best for the industry and for their environment—both. But that's overlooked. Nobody seems to care about that. They paint us as horrible ogres that are trying to destroy the ecology of the world.

What has happened there is they've made it so difficult to operate now, and we have all seen this. The majors in the United States are gone. They've left. They've gone to Russia, Indonesia, South America, everywhere. It's no longer economically feasible for them to operate in the United States.

That's evident from their exodus; look at their annual reports from their corporations. The expenditures in the United States are down substantially. They've left the United States. And do we want a domestic industry? I don't know whether they care or not. Certainly Hazel O'Leary doesn't care. She told us all that we were dinosaurs and ought to all go get reeducated as welders, or something like that. The energy secretary making that statement? That's a pretty good indication of what the government thinks. It's not coming just from the lower levels, it comes from the very top—the secretary of energy.

Somebody has got to bring a parity to this. We are not the ogres that are trying to destroy anything. All we're doing is trying to make a living in our chosen field of endeavor and, obviously, we're proud of being able to produce a product that helps the United States and the people of the United States.

We're guilty as an industry of not having presented a better image. I defend that by saying this: We are competitors—highly competitive with each other—independent explorationists are. We compete for the same acreage. We compete for the same knowledge in geology, for the same product—the reserves under the ground. It's very difficult to get two or more—a unified front—to pose an image. A public relations image is something that we need to improve.

Difficult to unify

But, again, I'm defending us by saying we can't. It's very difficult to have a group of men who are highly competitive present a unified image. We were told we should do that. That's another thing Hazel O'Leary and many other people have said.

There is another undermining factor there. Many of the organizations that purport to represent the industry do not. The Natural Gas Association, I believe, is one. It's made up of four or five transportation companies. They're not in the oil business. They transport a product. It might as well be loaves of bread. It happens to be petroleum.

They're not in the oil business. They don't go out and find oil. They don't explore for it. They don't operate the wells and they don't produce it. They buy it from us and they transport it to an end user. Yet they go to Washington as a natural gas industry saying, "We represent the natural gas industry. We would like these things to be done for the natural gas industry." They are proposing aspects that help them as transporters but are directly detrimental to explorationists. Nobody in Washington seems to care to look behind that and see who's who in these different organizations that play that game up there.

I don't know how that can be done, other than some form of legislation. It's misrepresentation for them to do that. It's misrepresentation for a group of refiners to get together with a petroleum organization or some pseudonym and go to Washington and say they represent the explorationists and producers. They don't. They represent refiners. Again, they're competitors of ours. They purchase our product. They want to purchase it for as small a price as they can. They will make proposals that are detrimental to explorationists and producers, purporting to be making those proposals for the industry as a whole.

There are very few of the lobbyist institutions, if you want to call them that, that are in Washington that actually do represent producers. TIPRO—Texas Independent Producers and Royalties Owners Association—and IPAA—Independent Petroleum Association of America. Even so, some of those are made up of people more oriented to owning drilling rigs—drilling contractors—than exploration and production.

There has to be some way to identify these groups as to what they actually represent. And that's our problem. We have never been able to control it. We don't have the governmental authority to control that. That is another reason I think we have a distorted public image, because there are so many different philosophies coming out of the industry.

There's not that many different philosophies coming from independents who actually explore for and produce oil and gas. Even though we are competitors with each other, our philosophies are similar. We would all like to do away with the alternative minimum tax. We would all like to see the return of the depletion allowance—which is not going to happen. Those are things that would be central to any independent group that would sit down and ask us what we would like. These are things we'd like to have that would help us greatly. Do away with expensive regulation.

OPA '90

I want to back up a little bit a minute and talk a little bit more about the regulations imposed recently. Some time ago, there was a requirement that in order to explore for, produce, or transport hydrocarbons in the Gulf of Mexico, you had to post a bond. Many years ago, I believe it was something like a million dollars; then it went to $5 million, $10 million, and $35 million. Now they want to raise it to $150 million [under OPA '90].

One of these bright environmentalists now has said, "Well, actually that should apply to not only the Gulf but all navigable streams." According to the Natural Resource Code, a navigable stream is one that is thirty foot, I believe, from cutbank to cutbank; that which can traverse a canoe. That means virtually every watered canal in the domestic United States. If I have to lay a pipeline to one of my wells when I drill it and start producing it, which we all do, invariably there is going to be some creek somewhere that you have to cross. There's very few of them that are less than thirty feet from cutbank to cutbank.

Does that mean that we as domestic, onshore producers who never intend to go in the Gulf of Mexico because of the financial requirements, could find that imposed on us? That's what the environmentalists want, and that virtually can put us out of business.

There's only about five thousand of us in the United States today. There is a small percentage still doing well. We're not doing great guns; we're just surviving. Things like that put us under. There is no possible chance that I or any of the men I know could post $150 million bond.

Making a difference

I don't know that we will ever be able to produce sufficient amounts of crude oil to make up for our needs again. I don't think we can in the domestic United States. But we can replace all of the difference, at least a great portion of it, with natural gas. There is enough natural gas in the United States today that we can approach, if not completely achieve, self-sufficiency. But not as long as we're being attacked by government regulations and environmentalists and the IRS. Because at every turn, they are at us again. We seem to be the golden cow—tax cow.

When you have principally ten states that produce oil and gas—there's more, but only ten that have a substantial amount of oil and gas production—and forty states that don't, that's twenty senators that will vote for it and eighty that will vote against it, because their feelings are that prices of energy being low is beneficial to the consumer, to their constituents, and to their reelections. The numbers just don't work. When you start putting it into Congress and try to do something beneficial for the oil industry, if ten states vote for it, forty vote against it.

Unfortunately, when they do take a more lenient attitude is in times of absolute necessity—when we have an Arab oil embargo, a disruption in supplies. But even when that occurs, they come back and say, "We've got a disruption in supplies, you guys need to produce a lot more oil and gas to help us out in the United States to get us out of the lines and the shortages of gasoline and all the different aspects that energy provides."

If they brought back the 27.5 percent depletion allowance and said no more alternative minimum tax, you get 100 percent write-offs, no questions about your intangibles/tangibles; and they did everything that would be beneficial to us, it's not going to be an instant turn-around. It takes time.

From the time you say you're going to drill that well until you put it on line and start selling the product—if you can do it in a year, you're doing damn good. That's if you have everything ready to go. It can be a lot longer than that, especially when you have to go through seventeen different governmental agencies just to get a permit.

The opening of vast areas of reserves that we're not allowed to drill on today is essential. There's tremendous amounts of reserves that we're not allowed to even touch.

Hopefully, we won't ever come to a point again where there is a disruption of supplies, but if we do, the public's going to get mad at us because they're going to say, "We've allowed you these legislative concessions. It's been two months, why haven't there been any changes?" It's not going to happen that way. That, maybe again, is our failure to communicate to the public that it is very complicated.

It's not something that you just snap your fingers and it all starts happening in one second. My Canadian friends, some of them think we're the stupidest people in the world. They produce up to the boundary between the United States and Canada, and on the other side of it we can't produce because those are governmental reserves. They're draining our production and they look at us like we're the biggest bunch of dunces that ever lived. We are for allowing them to do that, and as long as they're allowed to do it, they will. Then again, it's counterproductive. That doesn't help the United States. The environmentalists say, "No, you can't drill in that area." Well, if the oil is under the ground and the Canadians are taking it—they're taking it in Canada and selling it back to us—our own oil. It's absolutely, colossally stupid to allow things like that to happen.

Intangible costs

Again, thinking about intangible drilling costs, one of my pet peeves is that whoever determined what is and what is not intangible had to be under the influence of something, because they weren't thinking clearly. Traditionally, we have had to set surface pipe to protect surface waters—which is right, we should, and we're glad to do it. That pipe is set in the ground and is never removed—ever. Yet it is not an intangible drilling cost and no one has ever been able to explain that to me.

Why do you have to capitalize that? Why is it not part of the cost of drilling a well, just like the cost of paying the contractor? You put it in the ground. It never comes out. Nobody ever intended it to come out. It would be against the law to take it out. But you still have to treat it as a tangible, not an intangible.

There has to be some sort of logical, reasonable interpretation of that. Not just because somebody said, "That's the way it is and, the hell with you, that's the way we're going to do it." If it's wrong, it's wrong. They shouldn't do it.

Futures

Another pet peeve to me is the industry that has sprung up that trades in oil and gas futures. That's the most bogus, ridiculous thing I've ever heard of, and it is directly detrimental to the industry.

If the price of the futures is pulled down, it pulls our price down—our contract field price. There have been trading days that the equivalent total of barrels traded in one day was in excess of the total production of the United States in a year. The point I'm making is they are trading something that doesn't exist.

If they shut it off during the middle of the day—if at eleven o'clock they said, "Stop. Everybody fill your contracts—complete your contracts with wet barrels."—there would be lines in the windows on Wall Street for people to jump out of because they couldn't do it. It's a fantasy. You cannot have a commodity traded 365 times as much as exists in one day. It's an unreal situation.

When they draw the price down because of what they term "technical corrections," that doesn't have anything to do with the oil business. We didn't technically correct anything. We're still producing the same oil that we produced last month and the month before in the same way. Yet all of a sudden they have a technical correction. That just means they played too high. They bet too much. So they had to bring it back down. They bring our price down. The minute that happens, the posted prices for crude come down. It has nothing to do with the product. It's just that a bunch of people on Wall Street decided they wanted the price to come down.

It's very detrimental to us and leaves us at the mercy of the traders. It adds to the instability of the price. Consider this: if the price of oil and natural gas remained the same for one year—without any price fluctuation—those folks would be broke. Every one of them. They'd be out of business, because they make money on variations of price or fluctuations. The more volatility it has, the more money they make. If the price of natural gas sat at $1.75 per Mcf for a year and oil sat at $17.50 per barrel, none of those offices would be open, because they would not make one penny.

So they look for instability. They're again directly opposed to what we look for. We want price stability to help us find more oil and gas. They want price instability to help them make more money on the product that we produce. It's very counterproductive to us. We're looking at stability to be able to tell someone, "This is how many days it's going to take—600 days or 400 days—to get your money back." If you don't know what the price is going to be, it's very difficult to arrive at a recovery point.

Weakened infrastructure

Another serious problem is that the infrastructure of the industry is being severely damaged. That's probably the greatest danger we face. In the past, wherever you wanted to drill a well—certainly in Texas, Louisiana—if you needed some oil-field supplies, there was an oil-field supply store.

That's something you have to consider now—how far away are the supplies, because it's downsized so much. So many places closed. The service industry—all the services, not just supplies but logging, drilling fluids—has been greatly downsized. We've lost hundreds of thousands of employees in the last twelve, fourteen years. These were not people who just in a quirk of the moment decided they wanted to take a whirl in the oil business.

These are educated people that were highly trained to be in this industry. Geologists, geophysicists, petroleum engineers, mud engineers—people who spent their lifetime being educated to do these things. And they're gone. They couldn't make a living. They had to make it somewhere else. They are no longer in the oil business.

It's difficult to find qualified people, where it was not difficult in the past. Drilling contractors have to work to keep qualified hands on their rigs—the roughnecks and the drillers. When you start eliminating the equipment, each rig that's gone loses three crews. About four hundred jobs are created for every rig that works. So every rig that goes down, there's a whole lot of people that don't make a living because of it.

By the turn of the century, I think there will be a higher and probably more stable price for crude oil and natural gas. By more stable, it's likely to be an upward trend. It's not going to be a skyrocketing upward trend, but it's going to increase. I believe $22, $25 is probably a reasonable price expectation for crude oil, and $2.50 for gas—in that neighborhood. We could live with that. We could make a living doing that. But not if it goes from $25 to $10 and back up to $25—that just kills us.

More downsizing possible

I'm certain the role of the independent will be the same as it's always been, although ten years ago there were double or triple the number today [5,000]. By the turn of the century, I could see it being half again that small, down around 2,000, 2,500. Without allowing us more flexibility, it will be a greatly downsized industry.

We explore for and drill 85 percent of the newly discovered wells—have traditionally done that since the beginning of the industry, and still do. The majors seldom drill exploratory wells, preferring to let us take the risk and then buy it from us. It's simple business sense.

It's easier for them to justify to their shareholders the acquisition costs and the development costs of a new field they have acquired from an independent than it is to justify the loss of a dry hole. They'd rather us go take the loss of the dry hole and then come say, "Okay, I'll buy you out. I'll buy that field from you," and then they put up the money to go ahead and build the refineries and/or drill oil development wells or whatever. They can do it more in an accounting manner than a risk manner. The risk is there, but they eliminate the big risk. That's traditionally been the role of the independent, and it still will be. Perhaps this is contrary to the opinions of others in the industry, but I don't think so if they consider it. It will be more so because there is less activity by the majors.

The majors are moving out. They're not going to drill the new wells. They've pretty much put the world on notice. They're going to drill in friendlier environments. So they're not going to spend the money. Not just because of the risk, but they don't want the environmental problems. The problems are horrendous when you have to clean up these things. They don't want you just to tear the tanks down and plant the grass back. They want you to go down so many feet, and be sure that it's not contaminated soil, and have it cleaned to a more natural state and replaced. It's incredible how much it costs.

That's another reason many of the majors have sold fields and production facilities. They want to get out from that environmental problem and turn it over to somebody else. So they're willing to sell domestic production, maybe at a pretty good price, because they're looking down the road at what it's going to cost to plug and abandon those wells and clean up the mess left behind. They'll sell it to you for a fairly decent price, because they know you're going to have to look at that environmental problem.

False image

I don't know why the industry receives such attention. Maybe it's because of the historically incorrect vision that we're held in their eyes. They—many of the public and the people in the government—think that we're a bunch of wild-assed cowboys. You know, like the Beverly Hillbillies. You go shoot a hole in the ground with a rifle and it comes bubbling up, and then you go spend all the money in the world. There were colorful men who made a lot of money and did what the hell they wanted to with it. Those days have been gone a long time. I don't know that there ever were men or explorationists and production people who were as wild and woolly as they were portrayed to be. But that's not true now.

The industry is made up of professionals who are dedicated to what they're doing. We may be paying the price for a past caricature of us, but I wish we weren't. I don't know how to change that, other than we'd all try to live our lives in a very normal, productive manner.

George P. Mitchell

chairman of Mitchell Energy & Development Corporation

George Mitchell relaxes in his office suite high up in the Texas Commerce Bank building in downtown Houston. If you look far enough to the south, you might spot the outskirts of Galveston, Mitchell's hometown, where he has taken an active role in urban revitalization. If you look far enough to the north, perhaps you will see The Woodlands, the thriving suburban community planned and built by this legendary oilman and entrepreneur.

Mitchell is a living example of the American success story. The son of Greek immigrants, he sold stationery and bused tables while earning a B.S. degree in petroleum engineering, with an emphasis in geology, from Texas A&M University. After World War II, he and his older brother Johnny began a small wildcatting firm. "I did the geology and engineering, and Johnny would go down and sell the deals," he recalls.

On a tip, they discovered a huge natural gas field. In 1972 the company went public.

Today, Mitchell Energy & Development Corporation is one of the nation's largest producers of natural gas and natural gas liquids, and a major real estate developer in the Houston-Galveston region. Mitchell and his wife, Cynthia, the parents of ten children, are actively involved in the rejuvenation of Galveston's historic Strand District, and for the past twenty years they have been building The Woodlands community, which will ultimately house some 150,000 people less than thirty miles from downtown Houston.

Now in his seventies, Mitchell is far from considering retirement. "A lot of my colleagues want to turn sixty-five and get out on the golf course," he says. "I just don't understand that."

*T*he oil and gas industry has been so volatile in recent years that it's somewhat difficult to peg a single element that is the key to the future of the petroleum business. Taking a look at the big picture, it seems clear to me that the nation will be far more dependent on oil from the Middle East than it is today, so one primary concern will be getting the American public—and our elected representatives—to realize and acknowledge the importance of the petroleum industry in our daily lives. We can't continue to stick our heads in sand, ostrichlike, and hope that the problem goes away.

The problem with our dependence on imported oil is that three-quarters of the world's reserves are in Saudi Arabia and Kuwait and the other countries on that little peninsula. If all of the reserves in that little segment of the Mideast were divided among thirty companies around the world, then there wouldn't be a problem. But suppose today there was a serious problem with Saudi Arabia. How would we meet our energy needs? When Russia really starts producing in about fifteen years, and it is able to export oil, that situation might not be so bad. But until then, we are in a very serious security problem, and it has not been solved by any means.

Last year [1994], we passed the 50 percent dependence level for foreign oil, a benchmark that former Senator Lloyd Bentsen identified as the "peril point" for national security when sponsoring legislation to establish a floor price for crude oil. Unfortunately, that effort failed, as has every other attempt to develop an energy policy that gives us some control over our own destiny. For this country to formulate a sound energy policy, we need to answer one question: How do we decrease imports?

We wouldn't be having these energy problems if they had passed an oil import fee. I still favor an oil import fee, but nothing's going to happen from our elected officials on that. With the free-trade issues and treaties already in place, it gets more difficult to institute floor-price policies. What we need now is more tax incentives, such as tax relief on stripper gas wells. We need to get rid of the alternative minimum tax for companies trying to find oil and gas. That's what I think has to be done.

The government's role should be to develop an energy policy that makes the nation secure, but they need to work with the business world because they can't do it themselves. Nearly everything the government fools around with is a disaster, such as oil and gas legislation and the strategic petroleum reserve. The cost of imported oil is probably in the neighborhood of $100 a barrel when you figure in all the costs we have to pay to ensure its delivery.

What do I see in the future? In this Congress and in state offices, I'll expect to see a lot of improvement when it comes to working with the environment, such as fewer restrictive operating regulations. Environmental issues could make it so difficult and costly to drill and produce oil and gas in the United States that companies would no longer compete here.

I believe that the oil and gas industry deserves more attention from the government than other industries, because it is much more important than steel and autos and others. While the number of jobs the energy business provides may be fewer, oil and gas is essential to national security.

Unfortunately, any attempt to develop an energy policy that gives us some control over our own destiny has failed. Someday, the chickens will come home to roost, but until then, I do not expect Washington to take any steps to help preserve our domestic oil and gas industry. That means we'll see continued erosion in America's oil and gas productive ability, continued restructuring of the industry, and a continuing exodus of independents and majors alike as they seek their fortunes on foreign shores.

The natural gas solution

For this country to develop a sound energy policy, our national interest dictates that importing more than half of our daily supplies of oil is a problem. It's a security risk, and we must do something about it. And what we must do is use the natural gas that lies underneath our feet. It's waiting there for us, but we have to drill for it.

Naturally, I think that natural gas is extremely important for this nation, because it can substitute for much of the oil we import from other countries. The very dangerous level of 50 percent dependence on imported oil could and should be relieved over a ten-year period if a massive effort is made to encourage the increased use of our country's natural gas resources. This should be done, because it's a better fuel than anything else. Natural gas has fewer CO_2 problems than other fuels, and there's no acid rain to deal with. But the industry has to get politicians and environmental groups involved. And that will happen when we can show them that we're doing a better job of cleaning up our own backyard. Now, there's too much restriction mandated by Congress, and their hands-on approach should be replaced at the state level or by the industry itself.

We have to conduct more exploration in our own country. We're not replacing the reserves we produce in the U.S. each year. In the last four years [since 1990], we've replaced only 85 to 95 percent of the gas reserves we've produced. We can't keep that up forever. Gas from Canada can supplement our needs for a while, and maybe that's okay. We should work on the North American continent, but we really need to be sure that we use all of the domestic gas we can so that we offset our dependence on foreign oil.

Right now, the price for gas is a little low, but it will swing the other way when the demand for it increases. Companies can't explore for gas when it's priced at only $1.50 per thousand cubic feet. It takes $2, $2.50, or $3 an Mcf to fund an exploration program in this country to replace imported oil.

Techniques and technology

Another key issue for the industry is technology. We have to become more efficient using the latest technology, which I think is happening, albeit quite slowly. We can drill wells in half the time it took just ten years ago, but costs are going up, such as the price for tubing.

The newer independents and younger geologists understand the importance of the latest exploration and production techniques, and that bodes well for our industry. Most of the engineers also understand this, but I suspect that the ones who have been around here for twenty-five or thirty years haven't gotten into it.

The people in the industry must be reeducated and retrained to use this new technology. This is also very important, because at universities today enrollments are down about 80 percent for petroleum courses—whether it's geology or engineering. That level is abysmal, and that may be the biggest challenge our industry faces. We will be facing a shrinking pool of available talent in the years ahead.

Our industry's in a tough situation. We're going to lose the drive this nation has had for one hundred years if we don't focus on our energy needs. Our world leadership position in petroleum technology is at stake.

Trends

I think independents will become stronger. The majors will go overseas because they can't compete in the U.S., with its controversial environmental issues and the negative impact of costly operating regulations. With bigger discovery opportunities available across our borders, the majors can do better elsewhere. Deepwater offshore programs are an example of the big fields where they participate.

Independents will do more here, more in the shallow waters, and in some deep water. The independents have to get smarter. We have to get financing, but we can't get financing when the price of gas is too cheap. In the future, there will be fewer independents, but they will operate more intelligently, more efficiently, and they will be leaders. They'll have a role in supplying gas.

My company will stick it out right here in the United States, where we've carved out a niche that has kept us profitable year after year, even through the worst of the energy downturn. With more and more companies going overseas or out of business, additional opportunities may open up for Mitchell Energy, and we want to be in a position to take advantage of them. Like other companies, we are restructuring, disposing of non-core assets, and focusing on those areas that provide the greatest return on investment.

I think majors and independents are going to start doing more combined research. The majors are cooperating more among themselves. Up to about five years ago, they would never talk to each other on research or anything else, because of the threat of antitrust lawsuits. Now, that's been broken down, and they're actually partnering on a lot of research projects. That was unheard of years ago. Nevertheless, the majors will continue to move overseas, and with them, the technological edge the U.S. has historically held.

We're not doing the research in this country that used to be done among the majors. We've all cut back. Furthermore, now there are very strong research areas in Britain, the Netherlands, and France. For the last seventy or eighty years, all the major improvements were made in this country, with a few exceptions, such as the original seismic work from Germany and Schlumberger's work in France. But all the technological improvements, the drilling, completion, logging, everything, was done here.

Financially, the industry is still a good investment for people of capital, particularly if you get natural gas, which is the fuel of the future. In twenty-five years, natural gas will be lacing the world just like oil is today. Geologically, there is much more natural gas, so it will be easier to find than oil. We've got an eighty-year reserve of gas in this country, but it takes new wells and pipelines to connect these reserves and transport them.

I think that the potential worldwide is going to be natural gas, and that bodes well for this country because the environmental impact with gas is so much better than oil or coal.

We're still building pipelines, we're still drilling wells, and we'll be getting gas from Canada. Electrical power generation will be the next revolution. Those generators will use more gas. When the utilities monopolies break apart in the next ten to fifteen years, the future of natural gas will be even brighter. A very small amount of natural gas is used today to power automobiles, but as the mileage in those cars improves, with a combined program of gasoline and gas or some other technology, it will become more economical and more widespread.

People often talk about alternatives to natural gas, but I don't see one in the immediate future. Solar, fusion, hydro, and thermal power are too expensive right now. I think those sources right now come in at about $40 to $50 per barrel of oil equivalent. So they still have a long way to go. But we should do the research and investigate them.

In this country, I think more offshore areas should be opened up for exploration. Even the ANWR should be opened in ten years or so. Today's energy companies are able to explore and drill for oil and gas without harming the environment. I would say that offshore California, offshore Texas, Louisiana, and the East Coast could provide ample opportunities for the country to add to its oil and gas reserves.

Service Companies

ROBERT E. ROSE

president and CEO of Diamond Offshore Drilling Inc.

Robert E. Rose has been involved in the drilling industry since 1961. From 1964 to 1978, he held a number of positions around the world, including in the North Sea, Africa, and South and Central America. He began his career with Diamond M. Corporation, a wholly owned subsidiary of Loew's Corporation, in July 1979 and was named president and CEO in 1986.

In January 1992, Rose was instrumental in the acquisition by Diamond M of Odeco Drilling Inc., which created one of the world's largest offshore drilling companies. He holds an MBA from Southern Methodist University and is an influential member of many organizations, including the International Association of Drilling Contractors, for which he served as chairman in 1994.

Interviewed in his west Houston office, Rose talks of the need for the drilling industry to further consolidate and describes the regulatory and taxation frustrations the domestic industry continues to endure. Whatever problems beset the industry, it is clear that Rose will be among the survivors.

I believe that there are three major issues facing the industry today. One is the onerous regulatory environment which we are required to operate in unnecessarily.

Two is the fact that this industry is probably the most heavily taxed—in addition to being the most heavily regulated—industry in the United States.

And, lastly, is the lack of access to federal lands which we believe have promise in terms of oil and gas exploration. These three issues are the ones that you'd hear repeated most frequently by people in the industry.

Let's talk about overregulation. One example is the permitting process that is required for our industry to drill a well. If we were to stack on top of this table the documents that are necessary to be filed in advance of the opportunity to go out there and drill, that stack would be at least eighteen inches high. In some cases, in more environmentally sensitive areas, it could be as much as two feet high.

Mostly, these are reports that are filed with a number of the regulatory agencies, such as the Minerals Management Service, the Environmental Protection Agency, the Department of Interior, and the Coast Guard. Additionally, Congress enacted OPA '90—the Oil Pollution Act of 1990. This was supposed to be a "be-all, save-all" in terms of environmental spills in the federal Outer Continental Shelf. It was a knee-jerk reaction to a transportation accident in which the *Exxon Valdez* ran aground and spilled crude oil.

Congress's infinite wisdom enacted OPA '90, which requires the companies, regardless of size and regardless of risk, to demonstrate $150 million of financial ability prior to operating in any navigable waters. That doesn't just mean the Gulf of Mexico. In its most strict interpretation, that could mean rivers or lakes. Theoretically, a marina that's selling gasoline for an outboard motor must qualify for having $150 million worth of financial responsibility. That's an extreme.

Offshore on the OCS, where the perceived risk is the greatest, the largest spill that we've ever had in this industry cost about $35 million to clean up. It seems a little ridiculous to require $150 million worth of financial responsibility when, in our entire history, we've never had a spill (in terms of oil and gas exploration) that would even come close to that amount.

Those of us in the industry really resent the fact that our superior safety and environmental records are underappreciated. Focusing on the record of the drillers—and the exploration and production industry in general—our record is extremely impressive, in terms of safety and environmental protection.

During the twenty-two-year period of the OCS—which starts in 1971, and the last year for which we have data is 1992—there have been 21,715 well starts offshore. During that time frame, we have produced 7.387 billion barrels of oil. In that entire period, the industry has spilled from oil and gas activity 999 barrels. Now that's the record that we put forward in defense of having additional onerous requirements on us for pollution liability.

How strategic?

The government has decided, either rightly or wrongly, that our industry is no longer a strategic one. However, I think that our industry is important to the United States and, therefore, the role of government should be more as a partner with the oil and gas industry, as opposed to being an adversary.

We have seen throughout even some of our Republican administrations the gradual dismantling of our industry because of the encouragement of low-priced foreign oil. As a result, the oil and gas industry has lost more jobs than the automobile, textile, and steel industries combined.

Yet it's almost gone unnoticed in Washington. In terms of the overall strategy of our government, their philosophy has always been cheap oil is best for the United States, even though it is dismantling a domestic

industry. This philosophy requires us to rely more upon these politically unstable sources of energy.

As we all know, we had to go fight a war in order to keep the supply
lines of fuel open. Cheap foreign imports and send in the Marines. I
submit to you that it is not in the best national interest of this country
long term to rely solely upon that. It seems to me that if oil is worth losing American lives for overseas, it's worth drilling for domestically.

The federal government should take a much more balanced approach
to the industry in terms of allowing it to operate on the same playing
ground as everyone else. This is not a free market. It never has been a
free market. But, today, because of the onerous regulations, excessive
taxation, and lack of access, our own government is driving us out of
this country to seek opportunities abroad.

Propping up the FSU

I think it's particularly ironic that at the same time our government is
doing that to our domestic oil and gas industry, it's talking about aid
packages to the former Soviet Union to assist them in propping up and
supporting their oil and gas industry. Let me say that doesn't sit well in
Houston, Texas.

I could envision a partnership much like the Japanese have. I think
the government should be supportive of the effort. It should foster and
assist in technology, and help the industry in terms of looking at opportunities overseas.

If it is in the best national interest of the United States to assist the
Soviet Union, then why not do it with American technology and American companies? Why do it with American dollars? Why not assist the
U.S. petroleum industry by offering those services and by giving us a
competitive advantage in those countries like Japan, France, England,
and Italy will do?

But, instead, the U.S. government will end up, in all probability,
sending billions of dollars of aid, which will be spent to a large extent on
other countries' technologies and services.

The first thing I think that should be done is for the government to
really take a careful look at this industry and determine whether it is in

the best national interest of our country to be so dependent upon oil supplies from very politically unstable reaches of the world. If the answer is "yes," then so be it. But I have to believe that given the history of that region, that it is not in the best interest for us to blindly rely upon 50 percent of our energy sources coming from the Arabian/ Persian Gulf area.

I think the government can take more of a role in trying to provide a proper balance with environmental issues. I'm talking about balance. Let the American public know that *Exxon Valdez* is not related to drilling a well in the Gulf of Mexico. Those are totally different issues. And yet, I have seen instances such as when they are out there holding hearings on opening up the OCS in California, the environmentalists run all these scare tactics by showing photographs of burning drilling rigs, of *Exxon Valdez* and all the animals that were killed as a result of crude. This is a distortion, it is not the truth, and it is not in the best interest of this country not to explore for oil and gas domestically.

We, as an industry, resent it very much when one of our own, such as President George Bush during the peak of the Persian Gulf crisis, was calling upon countries to step up their production to alleviate any possible world oil shortage. This was at a time when we here in the United States could not drill anywhere offshore other than about four states and the Gulf of Mexico. Chevron, which had already drilled the wells which have the production capabilities in Point Arguello, California, could not produce it because they could not get the permits to bring the oil ashore.

That, to me, is totally unacceptable to the American public. I think if they understood those issues, they would have a much more balanced view of the industry. I believe if the public were to hear the arguments clearly and fairly in congressional hearings, where these facts were known to the public, it would be difficult for them to say that we need to continue to have over 50 percent of our deficit buying foreign oil when we can produce it safely here in the United States.

I think the public would see that it is no longer in the best interest of this country to continue to put off-limits all of these potential assets that belong to all of us, not just the people in California. Those are federal assets that belong to everybody in the United States.

I don't think the public realizes that there is no cheap foreign crude. That crude has been estimated to cost, in addition to costing American lives to keep the supply coming, as much as $90 a barrel in our defense effort to keep the Persian Gulf open so that oil can come here at $20 a barrel.

If all of the costs were known of how the public is subsidizing foreign oil, I think the case could be strongly made that we need to develop our domestic resources. We can never domestically produce all of the oil this country needs. But we ought to be producing all of the oil we are capable of producing domestically.

Focusing abroad

In simple terms, the days are numbered. If we're continually denied access to promising lands, at some point in the future, the Gulf of Mexico—which is the only market we can look at offshore—is going to be exhausted. We're going to have done all the exploration out there that we can do, and therefore the efforts are going to be largely international in terms of seeking oil and gas opportunities.

Initially it was the majors who moved overseas, out of the Gulf of Mexico, leaving behind a lot of opportunity for independents. Today, most of the wells drilled in the Gulf of Mexico are by independents. More and more, independents now are looking overseas, particularly with requirements such as OPA '90 that may put them out of the business here. That may be the only option they have.

With our oil customers domestically limited in the opportunities they have to drill wells, there are fewer wells drilled here in the United States. Consequently, the majority of our fleet is located internationally. We have a large presence in the Gulf of Mexico, but more than half of our rigs are located in foreign waters. It's certainly a major change from fifteen years ago. This has been a process that has been evolving as opportunities have been presented overseas. More and more countries are offering incentives to our oil company customers, versus the disincentives that are here.

The only thing the United States has to offer a company, versus another foreign country, is political stability, and that's not an insignificant thing. But that's really the only thing an oil company can put in the plus column when they are looking at drilling here versus drilling in South America, West Africa, or Southeast Asia. Everything else about the United States is in the minus column.

In many countries, companies are offered relief from royalty until payback of their investment. In some cases, the foreign countries are partners and actually put up risk capital with the oil companies.

In many cases, corporate taxes are waived until a certain level of profitability is achieved. Acreage is granted at minimal cost, as opposed to the expensive bidding processes. In many countries—Australia and the U.K., in particular—the way the work is granted over there, which certainly warms the heart of drillers, is done on a work-commitment basis.

If I want to drill on a particular concession, I bid a work program which says that during the first year, I'll do so much seismic work and the second year I'll drill so many wells. Whoever bids the most aggressive program is the one who's awarded that lease.

That generates a tremendous amount of economic activity. It's not like here, where you bid a certain dollar amount. The government gets the dollar amount you bid and you may or may not drill it. And if you don't drill it, you just forfeit it. These other countries, because of the way they offer their concessions, generate a lot of economic activity in running seismic, drilling wells, hopefully making discoveries, building platforms, and putting them on production.

Those are some of the reasons why people look overseas.

High-cost producer

The U.S. is a very high-cost producer because of the environmental concerns and because of our salary structure. Because of all the safety and environmental issues and government taxes, it costs a lot more per barrel of oil to produce here than it does in some of these foreign countries. Obviously, everybody being as profit-motivated as they possibly can be, will look for the cheapest source of hydrocarbons.

Part of the problem is the lack of access in some of the more promising areas where it might be low cost to produce. Since we're limited pretty much to the Gulf of Mexico, we find ourselves having to move out in deeper and deeper water in search of hydrocarbons domestically.

Therefore, the deeper the water, the more expensive the whole process is. This is in contrast to Saudi Arabia, where you can punch a hole in the desert and produce a barrel of oil for $1.50 a barrel. We are forced to move out there in deeper and deeper water, and produce that barrel of oil for $12.

Those are a lot of the problems. That $12 comes from the environmental safety and the taxation. So, if you can go overseas and produce it for less money, then that's where you're going to spend your investment. And as the price of oil continues to languish around the $18, $19 level, companies can't economically produce $12 or $13 oil here in the Gulf of Mexico. They've got to go find another source and produce it cheaper.

I think the entire focus of the industry is that you'll have a domestic gas business and maybe the remnants of an oil business. But the oil business will be an international one. The economics just tell you that's the way it has to be, unless there's some significant geopolitical change in the world.

Double-edged sword

Today's advanced technology is basically requiring that we drill fewer wells than we have in the past. Part of that is because of technology and part of it is because the leasing program has been changed. Today, you can optimize the land that you're leasing and drill fewer wells to either condemn or prove that particular area of the Gulf of Mexico. That's beneficial and positive in terms of the industry because it improves the economics, but it does require fewer wells to be drilled. So for the drilling companies, that's a negative.

Two technologies, I think, are two-edged swords. One is the 3-D seismic. Because of 3-D seismic, a lot of locations are being condemned by seismic that would have been drilled before. But the flip side of that is a lot of wells are being drilled because of 3-D that would not have been drilled before, because 2-D would not have shown these particular potential producing zones. I think, on balance, it is a positive power to our oil company customers by improving the economics.

Second is horizontal drilling. We can now tap into a number of reservoirs with one well bore, whereas before it used to require multiple well bores. Again, it improves the economics for the oil companies and requires fewer wells to be drilled. Hopefully, on balance to the industry, that's positive.

Technology, very definitely, is making us more efficient and more economic. Among future advances, I think we'll continue to use downhole intelligence. We'll continue to try to improve upon the measurement and logging while drilling. We'll have steerable motors where we'll be able to really drive the bit where we want it to go and know exactly what it's going through. A lot of people are now talking about 4-D seismic. If it is as much a benefit as 3-D has been over 2-D, then it's going to be an exciting technology.

Pursuit of capital

This has changed in a number of ways. Back in the late '70s and early '80s, where there was a very high tax bracket, you were able to raise capital on a limited partnership basis where people were able to basically spend tax dollars to drill wells and build rigs. Now that we have a maximum tax level, those funds aren't available to the oil and gas companies.

What you're finding is that people have to be a lot more successful in their efforts to access equity or debt capital for the oil and gas companies. Now, for the drilling companies, we've never really been in a posture where we can show an investor a definite return on his investment, so you tend to get more speculators in the business.

Our ability to raise capital, either debt or equity capital, is dependent upon the perception of the business more than the reality of the business. For example, in late '93 a number of companies that were private went public through an initial public offering, raising equity capital. A number of them went into the debt market and raised debt funds for the first time since the early '80s, when that market was available to them. But this was done on the basis of the perception that the economics of the industry were changing. They didn't, and, as a result, a lot of those companies did these IPOs that are now selling well below their initial offering price because the market didn't materialize.

Our business, for the drillers, is very much what our perceived earnings are going to be one or two years hence. Not what the realities of the business are today.

Consolidation needed

Unfortunately, there's not enough consolidation going on. It very definitely is the right thing for this industry to be doing. We're still, as a drilling industry, too fragmented, and we have too many competitors and too many drilling rigs out there. Consolidation needs to continue. If it wasn't for that blip in '93 where many companies went to the market and raised equity and debt capital, I think we'd have seen a tremendous number of consolidations.

They did not occur as a result of companies going to the market, basically getting their war chests full again, and therefore gaining extended life. That was a disservice to the industry on balance, because I believe a lot of those companies would not be here today had it not been for the ability to go out and raise capital.

I don't care which company you're with, but every CEO of every drilling company thinks he's absolutely the best and has the best company, and that's probably true. There's always a reluctance to turn over control of your assets to somebody else. I hope that sometime during the next five years, we see much more consolidation among the drilling companies. Hopefully, we'll become truly an industry that can manage its future, as opposed to continue to react to it.

Trimming down

I think the majors and independents are becoming more akin to one another. What we've seen happen with the majors—as they continue to try to squeeze cost out of their structure—is a downsizing, or whatever particular term you want to use. They're shedding a lot of the responsibilities that they traditionally have held, particularly in the design and execution of drilling wells. They're beginning to look more to service companies to provide that expertise.

Most every major company has already or plans to experiment with having wells drilled on a turnkey basis. That would have been unthought of five years ago. As part of streamlining, they're giving the responsibilities to others to perform for them. Typically, independent companies have always relied upon service companies to drill their wells, because usually independent oil companies are basically geologists with geological staffs who identify prospects and then hire somebody to do the engineering and drill the well.

The challenge for a drilling company is to be all to everyone. We must be able to offer that full range of services from that major oil company that wants to hire you on a day work[1] basis, which is traditional, to that independent that really wants you to do everything for him, to be his drilling department.

The challenge for us is to be able to do that profitably. Obviously, to gear up your staff to have that capability in-house is an expensive proposition, and you need to be compensated for that. A lot of companies are betting big bucks and are very aggressive in the turnkey business. The reason I'm not optimistic about it is because the drilling contractor assumes all of the risk for turnkey, and today that's extremely competitive business and the margins just aren't there.

I don't believe that it's in the best long-term interest of a drilling contractor to put all of his eggs in that turnkey basket. You have to have that capability. We have it. Most of the major drilling contractors have it. But we certainly aren't so aggressive in it that it's going to be turnkey or nothing else.

You hear a lot about partnerships and sharing. I think we'll develop something in between turnkey and day rate that is the optimal way to drill wells. We'll squeeze cost out of the structure, and it will be a much more cooperative environment than we've seen in the past. Many companies are moving in that direction. No one has hit upon the right combination yet, but certainly, as an industry, everybody is working hard to achieve that.

That's really what we have to do. You give me all the risk, I'm going to put a big margin on it for that risk. Now if you take the responsibility for the risk that you traditionally should take, and I take the risk for

---

1. Day work means that a producer takes the risk by paying a daily rate for the rig, regardless of how long a job might take. The operator would also pay for all the materials that go into the well. Under turnkey, the drilling company drills at a specified location to a specified depth for a lump-sum price. Day work is still dominant, but turnkey has grown in popularity the past two years as many operators downsized their companies, forcing them to outsource work they normally did in-house.

the things I should take, there's less of a tendency to put these large contingencies in the overall cost.

Changing of the guard

I think we've decided that the days of iron men and wooden derricks are over.[2] You've got to have executives who are more global in their thinking and are looking with one eye on the here and now but another eye cast on the future.

It requires a much more sophisticated businessman than in the past. A lot of the previous executives in this industry were entrepreneurs who invested and risked great sums of money in very risky investments, and were fortunate enough to have made a lot of money.

Today, as we become more and more oriented toward producing quarterly earnings for shareholders, you'll find people who don't have that risk profile in their psyche. They tend to be a lot less inclined to put large amounts at risk.

It's a shame in many ways, because that's really the kind of man who made this industry in the initial stages. But I think we have matured now, and it requires a business approach. The entrepreneur still has a place, albeit a smaller place, but I think largely the executives are going to be businessmen.

2. According to Rose, "In the early days of the industry, when they put up a derrick, it was a wooden derrick. They were iron men back in those days because they were sort of expendable—they were required to do anything and you didn't care too much about it. So, 'iron men and wooden derricks' is a phrase I have for the beginning of this business."

JAMES C. DAY

chairman, president, and CEO of Noble Drilling Corporation

Formed in 1921, Noble Drilling Corporation is among the most venerable off-shore drilling companies in the world.

James C. Day has worked in the energy sector for almost thirty years. He joined Noble in 1977 and was elected chairman in 1992, after having served as president and chief executive officer since 1984. Day is an economist by nature, with a degree in business administration from Phillips University in Tulsa, Oklahoma.

I interviewed Day twice. At the time of our first meeting, the drilling sector was still struggling to come out of a steep downturn. By our second meeting, the drilling industry was beginning to rebound, thanks to growing interest in natural gas and a determined effort led by Noble to remove some of the excess rigs from the sagging market.

Day combines a scholarly appearance with a neighborly demeanor. He is deeply concerned about the image of the industry and frustrated that the public and lawmakers still link environmental issues to drilling. Like his peers, Day appreciates the opportunity to talk about his business.

*R*elative to the oil industry, specifically, there are three vital issues we're concerned with: pricing for oil and natural gas, capital requirements, and personnel. First, we need stability in oil prices. It was announced that Saudi Arabia was going to stand pat on its current production levels, and in fact would entertain the possibility of a reduction in production to assure prices are maintained.

Obviously, that's a change in philosophy. And, if that change is more than just a near-term anomaly, then it's positive. Oil prices need to be in the high teens or low $20s for the industry to be relatively healthy, worldwide.

Domestically, the driver will be natural gas prices. It will be the product that, if you will, provides a renaissance in drilling activity. There are two schools of thought. There are those that say we're on the cusp of a demand/supply balance, that gas prices over the course of the next twenty-four months should stabilize, and that we should see some strengthening through the balance of the decade.

The other school of thought is that we will live with overcapacity, since we've become much more efficient in our drilling operations coupled with improved technologies. Clearly, the overriding issue will be the need for stability and appropriate pricing levels for oil and natural gas.

Let's assume that issue is resolved. The second and third issues that are facing the industry are capital and personnel. We've gone through the decade of the '80s, certainly on the service side of the business, where we've been living on yesterday's, if you will, savings. Clearly, there's a need for capital as the market improves. That issue is not going to be resolved anytime near term, because the industry is not earning an adequate return on investments.

Thirdly, it's been stated many times that we've lost three out of seven people in the industry since 1982–83. We anticipate some growth in the business, assuming that oil and natural gas prices stabilize or at the very least stay within a narrow pricing range. We're short of people, and trying to bring new blood into the business has been and will continue to be difficult. Those basically are our major challenges—but certainly not all of them.

Battling costs

As an industry, we must improve our overall operating efficiency with the current fleet, and at the same time, maintain the current fleet at the highest technological level. Capital costs in the drilling industry are very significant, and we must focus on driving down the drilling costs for our customers. Offshore assets, 300-foot jackups specifically, will cost between $50 million and $120 million to build. Day rates today aren't at levels to construct new assets, necessitating the utilization of older assets.

To accomplish meaningful operating efficiencies, we need more capital to upgrade assets. We need continued improvement in technology, and we need to improve the skills of our personnel as we become more automated on the rig. The severe downturn during the '80s has not allowed us to make significant strides.

You can still fund companies that have done quite well, even during difficult times over the last decade. Especially in certain international arenas, it's a good business to invest in. Even domestically, there have been opportunities to purchase production and drilling assets at reasonable prices, which provide acceptable returns.

Generally, we have very strong, very independent personalities in the drilling industry. While there have been very few in our sector that have really been able to achieve earnings, at the end of the day there has been an unwillingness by many in the drilling sector, if you will, to throw in the towel. Small companies, in my opinion, will not be able to survive tomorrow because of the ever-increasing cost of operations. The cost of conforming to new regulations and the cost of training for personnel will constrain small companies.

Reason for concern

I don't believe Washington or the general population in the United States has recognized that the energy industry is important and that we can't become ever-more dependent on foreign sources of oil. It should be frightening that we passed the 50 percent import level in 1994. There needs to be a very focused effort to educate the population in general, which should be conducted jointly with various industry associations.

A reasonable energy policy could be predicated on drilling inducements, such as tax credits, royalty holidays, etcetera. Given the environmental concern, we should try to encourage, through again tax credits, the use of natural gas in vehicles.

I don't believe that there needs to be any type of import fee. I think you would achieve the same results—that is, finding more energy domestically—by providing royalty holidays or tax breaks.

I prefer to see government play a minimal role. The oil and gas business is no different than any other industry sector in the past decade, in that we've had a great intrusion into our operations. OPA '90—the Oil Pollution Act of 1990—has placed significant new requirements on our sector, with the end result discouraging domestic drilling—just exactly the opposite of what should be occurring. That particular act was enacted in reaction to the *Valdez* oil spill. It really doesn't relate to the results that have actually been achieved: for example, less than 2 percent of all oil spills relate to offshore drilling.

Keep government out

I think government will be most effective when it does the least in any industry, not just ours. I think that what they could do, as I've indicated, would be to provide some tax relief or tax incentives to our industry. The probability of that is pretty remote.

Now I find that quite interesting too, because we're one of the few nations in the world that really penalizes the energy sector. You need to only look at the North Sea, where there have been inducements to drill. The British government was very interested in having a viable energy sector and viewed it as an important asset. For some reason in the United States, we—and I'm not speaking collectively—the legislators and many citizens have taken the position that they want a weak oil and gas industry, which in the end will not serve this country well.

We in the industry need to do a better job of communicating what we're about, what we do, how important we are to the economy in general, how many jobs are involved, what the industry does not only here but what it does overseas, and the fact that we're able to export our expertise is a strength for our economy.

We have some of the best technology in the world, and we have some of the very best-trained people in the world. We do need more attention. How you achieve that, I'm not sure. We've made several false starts in the industry to try to get more attention. The average citizen views the oil and gas industry on the same level as politicians, which isn't particularly gratifying.

Certainly since the mid 1800s, oil and gas has been a driver for many new technologies. It's provided the needed revenue to allow this country to grow and to expand. Cheap oil and natural gas have allowed this country to be competitive worldwide.

Our culture

I've been asked, "Why can't we wipe out the whole oil and gas industry? Japan doesn't need it and they are very competitive." The reality, of course, is that their entire economy is predicated on a different basis. They've been able to structure their economy based on the need to import fossil fuels.

Our economy has been developed in the reverse, so to assume we can wean the entire nation overnight would be impossible. The future of this country and the future of the oil and gas industry are inexorably linked, and they will be way into the next century.

I've got a middle daughter who came home a couple of years ago and said her teacher knew that I was in the oil business and had made several comments about how we don't need to drill in the United States; we can drill overseas and bring everything here. And we can save all our resources here.

Well, that's a wonderful argument until you reflect on the fact that you've got all these ships coming in here, and the *Valdez* is a prime example. You're going to have more and more chances of more and more *Valdez* incidents. To think you can shut down domestic drilling and haul everything in from overseas is absolutely ridiculous. But here's a teacher influencing a child. Our child is independent enough that she questioned it and raised some pretty good arguments. Then [*he adds with a chuckle*] she came home and questioned her dad, "Why can't we do that?"

I think that a strong economy and stable oil and gas prices go hand in hand. The oil and gas industry wants a vibrant economy in the United States. Stable oil and natural gas prices are a prerequisite for businesses to make reasonable forecasts. Unfortunately, in real terms, oil and natural gas prices are historically low. They're lower today than they were, when adjusted for inflation, ten years ago.

Industry's errors

There were valid environmental concerns in many areas of operations in the energy sector. There have probably been habits or processes in years past that were not acceptable. I think the industry has done a very good job of addressing many of those concerns. This industry is, in general, very environmentally conscious. Certainly Noble was very aggressive early on and focused on the impact of drilling on the environment.

I think, though, that the pendulum has swung the other way, whereby the environmentalists in general have overreacted. In the case of the ill-advised OPA '90 legislation, I think the government overreacted without really looking at the facts.

Certainly, we're able through natural gas to improve the environment in the United States. Yet, at the same time, while we're trying to encourage the use of natural gas, moratoriums and regulations constrain drilling both offshore in the United States and in certain areas on land. That whole concept is really a strange relationship.

You've got a product that can help the environment, yet environmentalists in many cases have taken the position that they're going to limit its use. It doesn't make any sense at all.

We need to develop a spirit of cooperation. We support various environmental groups, because we're just as interested in preserving the environment for our children and grandchildren as the environmentalists. Unfortunately, the leaders of the environmental sector have been unwilling to sit down and talk in a reasonable fashion on how to resolve various issues. They've been very strident and have not been willing to discuss the development of energy in the United States in a reasonable environmental way. They take the position that they're going to limit the development of energy.

There are many examples of our efforts. Different drilling fluids today are much more environmentally friendly. In certain offshore drilling applications, where you're catching rain that falls on the drilling rig and carrying it inland to process the fluids in order to assure that there's nothing that is emitted into the water, are good examples of cleaning up the environment. That's not just because it's required, it's because the operator that's operating our rig is concerned about what might happen.

For some of the best fishing offshore, go to any platform rig. You don't see boats tied up to a platform rig because they like the looks of it. It's because the fishing is so great. You create an artificial reef.

On land, where you are actually covering the location with plastic and making sure that all products are hauled away or deposited in a proper repository, are efforts that are underway right now that assure our drilling programs are much cleaner.

Political change

With the change in Washington, there's a possibility that there may be some offshore acreage that's opened up. I think it's going to be slow. But, again, when we request access we're going to have to show specific plans on how we plan to develop those acreages.

It's incongruous that one of the main issues that the Republicans are focusing on is the budget deficit. Yet we are unconcerned about our trade deficit, which is significantly affected by importation of oil.

This means simply that fewer dollars are circulated within this economy. It could be going to the Middle East, it could be going to South America, or it could be going to an African nation. In all instances, it means that it's dollars pulled out of this economy, not circulated in this economy, not spent on the infrastructure, not spent on education, not spent on new highways.

It's silly that we don't develop reserves domestically—not to say we can totally eliminate the importation of oil, but we can certainly moderate its growth.

Environmental studies should be conducted in sensitive areas to determine whether drilling can be conducted. Once a sound analysis has been completed, then a decision can be made. Certainly there are areas that are unique, but those are the exceptions rather than the rule. Broad areas offshore don't have, or shouldn't have, constraints as far as drilling is concerned.

In a perfect world, we should have opportunities to drill on the East Coast and on the West Coast. In both areas, drilling can be controlled. On the West Coast, there are oil reserves that could have a substantial impact on domestic production.

Of course, there is ANWR. Given the results that we've achieved environmentally on the North Slope, I see no reason that ANWR shouldn't be opened up.

We had such economic pressure on us in the mid '80s and early '90s to downsize, that many of the people that left the industry have gone to other sectors, such as construction. As activity has picked up over the last two years, we've had to go out and recruit new people and train them, and that takes time. It's been particularly severe in the service sector.

They could be encouraged back when they see sustained activity—and sustained activity is a function of sustained pricing—and we haven't had that yet. Fortunately, in the last two or three years we've had some strengthening in commodity pricing, but what we need is sustained pricing. I'm not suggesting we need mid-$20 oil. What I'm suggesting is that we can have a sustained oil price of $20 a barrel and you would see activity pick up. I would also suggest that if you saw sustained natural gas prices of $2–$2.25 per Mcf, that you would see activity pick up.

Realistically, there will be delays. We couldn't respond immediately, but we would respond. It would take some time to return assets to a good operating level. There will always be a debate as to how long it would take to respond—I think six months to a year.

Electric cars

Over time, oil and natural gas will be supplanted, but not in the near term. Mandates for electric cars, while interesting, will be a long time in coming. First of all, they're more expensive than their oil and gas counterpart.

What no one has talked about, and I find this particularly amusing, given the environmental sensitivity to oil and natural gas, is what happens to the batteries at the end of their useful life? What do you do with them? I would submit that, depending on the type of batteries they are using, there are going to be caustics with heavy metals involved. The question is, what are you going to do? Are we going to have a large mountain of batteries to waste away into the next century?

There are some questions with alternate sources of energy that haven't really been resolved yet. There's a lot of excitement with electric cars, but at the end of the day, one has to talk about the residual exposure. Ethanol has problems of its own as far as the environment is concerned.

So, there are some other things that may replace oil and gas, but we haven't thought through the impact of those replacements and what happens to the environment. What will happen to the residual products over the long term is what I'm worried about, not as much the near term. Obviously, as a country we need to look at alternate sources, but from an economic standpoint, it's a long way off before it makes sense.

Some of the petroleum companies have worked on this, such as with solar power. The problem, again, is a lot of this technology is decades away from really being proven, and it demands a significant amount of money. Our whole sector has suffered from a decline in earnings and R&D [research and development] is generally one of the first areas that suffers when you have declining earnings.

In the twenty-first century

Twenty years down the road, I think we're going to have a different country and industry. We're going to have a country, unless we take strong measures today, that's going to be dependent on 70 percent to 80 percent of its oil from overseas sources. We could have oil shocks.

We're going to have a future where it's going to be more difficult to plan.

Certainly, we will be allowing a sector of the economy to fall—one that has led the world with its technology. We're going to allow international interests to step in and take the lead in the oil and gas sector. We will find ourselves—as we found with other areas, such as the electronics business—to be supplanted.

Unless there is strong leadership, I'm concerned about our industry for the longer term. I'm talking about strong, cooperative, positive leadership in Washington and a cohesive industry voice—which we haven't really had. It's been pretty fragmented, but we need a cohesive industry voice that speaks for the oil and gas industry in a unified manner.

Historically, it has been a risky business. Obviously, you're never going to ascertain in most cases, unless technology improves dramatically, a geological theory without a drill bit. You're going to have to commit some risk dollars. So, it's difficult to know whether the risk will have any correlation to the reward, but that's the nature of our business. Individuals make commitments, and at the end of the day, hope it makes economic sense.

The reality is that the day a well begins to produce, it's depleting. An oil and gas company's alternatives are to either extend the life of a reservoir or to find a new reservoir. Our business is certainly not static.

ALTON J. McCREADY

former president of Schlumberger Well Services

The name Schlumberger is synonymous with electrical well logging, and has been ever since the technique was first developed by French physicist Conrad Schlumberger nearly seventy years ago.

Schlumberger first developed the theory of electrical well logging in 1911. He and his younger brother Marcel, a civil engineer, founded La Société de Prospection Électrique (SPE) in 1920. Seven years later, their people performed the first downhole electrical log on a 1,500-meter well in France. In 1929, Shell Oil Company ran the first electric log in the United States. Today, electric logging is a basic branch of geophysics and is considered as important to the search for oil and gas as the X-ray is to medicine.

Alton J. McCready worked for Schlumberger for thirty-four years. In 1959, after graduating from Louisiana Tech University with a degree in physics, McCready was hired by Schlumberger as a field engineer in Natchez, Mississippi. By 1992, he had worked his way up to president of Schlumberger Wireline & Testing, North America, which was later renamed Schlumberger Well Services, North America.

I met McCready in the summer of 1993. The oil-patch veteran was one of the most knowledgable, witty, and animated executives I had the pleasure of interviewing. He also provided one of the most compelling accounts of the vast changes caused by the massive restructuring of the services sector.

McCready's speech was highly animated, with many of his comments punctuated by a fast-pitched "huh?" "OK?" or "you know?" to emphasize his point. Sadly, this was one of his last interviews. McCready died in February 1994 after a short illness. He was only fifty-eight.

I've been in the business, what, thirty-four years? It was the first job that I ever had out of school. And it was just an interview process. I knew very little about oil and gas, although I was raised in north Louisiana, around the oil and gas business. My family was not in it. I knew a few people in it, but basically I started in an entry-level position as a field engineer with Schlumberger, and I've been with them ever since. Not because I had a driving desire to do something in the oil and gas business, but it looked like the most attractive job at the time.

The public needs to consider what's happened in North America in the downturn of the industry—actually since 1982 it's been a continuous downturn, although there's been a few flat spots all the way down to '86, and even up to 1992. The viability of the industry to respond, the ability for the industry to produce hydrocarbons in North America, is almost to the point that it's beginning to be hard to do it. The environmental issues are big. The political issues are big and, at the same time, you've had a reduced price of gas, and oil prices also have been, certainly at best, in the mid to low range.

There also needs to be an understanding of the number of jobs that this industry generates—500,000 jobs have been lost in the industry since about '82; the big part in the economy that we play. The ability of the U.S. to burn cheap energy is a big issue in being competitive around the world. The complacency of people in general about what position we really play is big-time disturbing. The other is, in a political environment, there are no tax incentives for risk-taking in the oil business anymore, which makes it very hard to develop, because it's a very capital-intensive business.

Days of the Seven Sisters

The misunderstanding about the industry goes all the way back to the Rockefellers, and it had to do with the major companies and their control of the domestic market at one time. Six or seven companies controlled basically the whole market. There's a total misunderstanding of the independents that operate in this country—the amount of dollars they put into the industry from an investment standpoint, the amount of reserves they find, and the amount of oil they produce.

Shows like "Dallas" and movies like *Giant* did a great disservice, because people don't realize that there's a tremendous amount of people required to produce hydrocarbons. Smart guys, well educated, the whole bit. Tough industry. Really a tough industry, and people don't realize that.

We haven't done a very good job in general as an industry in trying to come to the public with a strong single message, because it is a number of companies involved and a lot of independents involved—what's good for one is not necessarily good for the other. We're not like the auto business—they all build cars. We all produce oil, but in environments that are different, methodologies that are different, and driven by different reasons.

We haven't done a very good job in public relations. I don't believe the industry thought they had to for a long time. I don't think they much cared, which is too bad. Now it's at the point that, like I say, the environmental issues have come up, oil spills get huge publicity. The implication is that if you drill wells, you create problems in the environment. You look at the Arctic National Wildlife Refuge [ANWR], a big emotional deal for the general public today, but an absolute necessity for Alaska, as a state, to survive. And also, in the long term, to replace reserves.

The impact of environmental regulations for us hasn't been as direct as it has for the oil companies, although we're heavily involved in that. We certainly have had to do a lot in the areas of radiation, which we're dealing with, and explosives, fluid control at the well site, and this sort of thing. Understanding what the requirements are is all time consuming.

We've added a totally environmental group within Schlumberger to address all the varied problems that go with it. But how it affects us at the well site is directly coupled to what the oil companies have to go through. The big problem with the oil companies is that they spend an inordinate amount of money in trying to get permits for lease. They have to deal with state governments all over the country. Some are favorable and some are not.

I think that the states in general have found out that you can make money in the environmental business. They do a lot of things that are detrimental to developing their hydrocarbons as far as the state is concerned—with the idea not so much for the environment, but how do they make money on it? The oil companies have a tough time dealing with it, and what happens is that they finally give up and say to heck with it, it's not worth it. And it affects our business, because if they don't drill wells, we don't have a service business.

It certainly has made a tremendous impact on the independents, because they don't have the staff. They have to go out and hire people to deal with these environmental situations. It gets to the point sometime it's so burdensome that they just decide not to do it. I think that's the biggest single issue that's there.

If you have producing properties, for instance, in the waters of Louisiana or Texas or whatever it happens to be, and the wells have been producing for a long time, it's an economical situation. If you have to rework wells, and they cost you a lot of money because of environmental regulations that are imposed on you, it's an economical situation. They may back away from developing that property, because it costs so much money from the environmental side. So it has a tremendous impact. You should do things environmentally sound, but there are some ridiculous things out there that they're asking industry to do. And it is driven because the states can make money on it.

No argument

When you say "environment," it's like motherhood and apple pie. Everybody wants clean air. Everybody wants clean water. They want animals, fish, the whole bit. But there's a way around that, and some of the things that the environmentalists push is just absolutely not true. If you go offshore in the Gulf of Mexico today, on producing platforms there's fish all over the place. People fish all over that area. The disposal of drilling muds, these kind of things, if it is a detriment to the environment, fine. You'll find the oil companies will support that, there's no doubt about it. But when it's a regulation that doesn't make any sense, that's the problem. And they have a hard time fighting it.

There is room to coexist. A prime example is if you go to Prudhoe Bay, Alaska, and you look at the pipeline, look at the Prudhoe Bay field. It's clean and environmentally sound. People are very aware of the situation that's there and all the political implications that are involved. Not only that, they want to do a good job.

It has been a very nice example of what you can do and how you can coexist in an area. You go up, there's wildlife all over the place, the migrating herds and what have you, as far as I know, have not been affected. When you look at something like ANWR and the size of that area, and the number of wells it would take to develop the project, it's miniscule.

In this country, when you have imports that's going at 50 to 60 percent, the only way you're going to get it here is tankers—and there is always risk involved when you have tankers. You do the very best job you can, but you're going to have that problem. You need to develop the properties in the U.S. You know? You need to explore areas where you can replace reserves and continue to have the cheaper energy cost or whatever.

Offshore California—huge reserves, big emotional problem. They've had one problem with an oil spill that happened to be in Santa Barbara [1969], through the whole history of the industry, and it's a big emotional deal. But the California people want cheap gasoline. It can be developed. You have to get the permits, you have to build some trust. It's a show-me deal. But you can't make it so expensive and so risky that the oil companies are not willing to take that risk. They just say, "The heck with it. Not going to do it. I'll go somewhere else. Overseas."

Look at the environmental issue in North America today. Do you know that it's harder to get permits to drill in this country than it is in three-quarters of the rest of the world? It is absolutely incredible. And we consume more energy than anybody else.

It's our own industry that creates jobs, too. We talk about jobs in this country. We lost 500,000 jobs over the last few years, and nobody gives a damn. Now, we're talking about cutting the defense spending today, and we're going to fund programs to retrain those people to go to other industries. Five hundred thousand people leave the oil business and nobody gives a damn. Tough. And part of that is just because the image that we have—programs like "Dallas"—is either you're a Texas millionaire or you're a roughneck. There's nobody in between. But there's a whole industry in between. And a very vital industry to this country. It's amazing.

Restructuring

We've restructured along with everyone else. The service industry is technology-driven and we were heavily tied to drilling rigs, although we do a tremendous amount of work on the well after it's under production, or during the production phase. But as the drilling rigs go down, you have to react, and what you have to do is that you end up laying off people. You have to size your organization so that you can make money and the environment is dealt with. We've had to do that.

What you'll probably find is that the service industry reacts a lot faster probably than the oil companies themselves, because our lifeblood is money that we generate in the market today. It's not wells that have been drilled ten years ago. We have to react very quickly to that. We've done what you see in most of the trade magazines or whatever: we flattened the structure. We have less management. We continue that process. We're trying to put all of our product lines closer together. Although we want to remain autonomous in our product lines, we're in the process of sharing a lot of the common administrative services.

It cost Schlumberger a ton of money to let people go. We tried to have reasonable severance packages. We did a lot of the things that a number of companies did—early retirement programs. When that didn't take care of everyone, then we had to go to severance packages, outplacement, all of these kinds of things—help our employees as much as we could—so we spent a tremendous amount of money in downsizing. You had real estate that you owned that you no longer could use that had no value, that we had to write off the books. When you look at the whole process, when you look at the financial cost of what we've had to do in the last ten years, it's tremendous.

The other side is that as a company, we have always developed a relationship with our employees that is kind of a long-term, career-type job. Work that's fun. I won't say a family environment, but certainly one that tries to make the work process worthwhile because of the nature of this work, twenty-four hours a day, oil field, and what have you. It's been a close-knit group. But we've had to cut an awful lot of people that are friends. And it's a much deeper cut than we would have liked to have done.

It's been tough. Families located in small towns and what have you, with no prospects, and they end up without a job, you know. What do they do? It's been extremely difficult. Not only for us, but for everybody in this oil business, because we all operate in the same geographic area.

It's different being laid off in Houston, Texas, than it is in a small town somewhere. It has a tremendous impact on people. And it has hurt the image of the oil business because of the volatility that's there. People look now at the business in this downturn of ten years and say, "You know, I'm not sure I want to be in that kind of business." So that's hurt. It's all hurt.

We've got about 60 percent of our people that we had ten years ago. People-wise, you're talking in the range of 3,000, from about 9,000 in the U.S. I think most of the companies probably have cut about what they intend to cut. Now the major oil companies are still going through a restructuring and that will continue, depending on the properties that they find or—make money.

If we want to attract young people back into this industry, it's going to require a couple of things. Stability will be one. Secondly, the oil business is a fun business. It's a challenging business. It's not sitting behind the door all day in an office job on the computer. If you like hands on, a certain amount of risk, this sort of thing, the oil industry always has an attraction, regardless of the ups and downs and the volatility of the business. We're back hiring. We're hiring field engineers again, and we plan to continue to do that. As long as you make the job attractive you can get good people.

This industry runs in cycles of about ten to fifteen years. It's hard to tell exactly, but it's not unusual to have downturns. Just like real estate. It's like a lot of businesses, although ours has certainly been whipsawed in the last few years. But if the gas price remains reasonable in this country, if we don't have a complete upheaval in the Middle East, and oil prices remain fairly reasonable, then maybe we're at the bottom of this thing.

If we are, I think that you will see additional jobs generated within the industry. It'll be a different type of an industry than we had before, in the way we do business, but you'll attract the people. I know that we are today.

Back to reality

There's a certain amount of realism that's here today that probably wasn't during the buildup of the 1973–81 era, when you heard talk of $100-per-barrel oil. There's been a lot of tax laws changed, so you don't get tax breaks for investing in the industry. As far as where we go from now and what's the buildup from this point, I think people have restructured really down to the bare bones. I know that we have, or we feel that we have. The long-term prospect is that if we grow, we will grow only in the money-producing side of the business—the field. People that generate money.

The charter's changed for the major oil companies also, in that they had these huge staffs looking at technical deals and what have you. The relationships between ourselves and the major companies are going to be much different than in the past. We will begin to be much more involved in the technical side of their business and helping them manage that.

What's happened with the industry is that they're finding out that competition is fine, but competition for competition's sake is a costly situation. They're looking at more and more ways that you can partner, that you can trade off of each other's strengths.

There's a certain transition that's taking place. Instead of having big technical staffs in the oil companies that look at the logging process, for instance, they're beginning to remove some of the technical staffs and they're having to build some trust within the industry, between our two companies. But technology application is only going to be used where it makes sense—where there are gains in efficiencies and where it makes some money.

You're also going to be dealing with people within the companies now that are not as technically oriented as your company may be. So you've got to represent the product line that you are dealing with in a manner that is well understood within the oil company by business managers. Not technicians, not engineers, not scientists. Where they had those staffs that interfaced with the service companies before, that's changing.

Washington can help

The government can do some things that are smart from an environmental standpoint—open up areas that need to be developed with a coalition between the oil companies and the government; put in some tax programs that make some sense, so that people who invest in the oil have a chance to make some money or an opportunity to write off some taxes. I think that's going to be a problem. If you don't do some of those kind of things, my recommendation is leave it alone. Take as much regulation off of it as you possibly can, run it like any other business, in a very competitive mode, and let free enterprise take care of itself.

The movement overseas will have a direct effect on our industry. For us, we are fortunate in that we have always operated internationally. If the industry decides to move overseas, we will size the operation in North America so that we make some money, and we'll put our resources overseas. At Schlumberger, we're not going to lose anything from a technology standpoint. We apply it all around the world today.

Because of the downturn of the business, companies that are basically domestic will have to go overseas to make a living. They're going to find that in the long term, to be successful overseas, they have to operate there, and they'll end up building overseas.

If the industry continues to shrink in North America and grows overseas, it'll be exactly what you would imagine. The jobs will go overseas, therefore the technology will go overseas. In the past when you looked at an expert in the oil business, he came from North America. Today that's changing around the world.

Long-term prospects

I don't think it necessarily has to be an irreversible movement, as long as you make money. It may all shake out in the long term. I believe there was so much pessimism and gloom back a year ago, 1992, in the end of '91–92, that people were about willing to give up on the American industry. It was an overreaction. There's still a tremendous amount of oil and gas to be found in this country. I think the infrastructure will be there.

It will mean that the oil companies have to make money, which means the service companies then can make some money. But if the oil companies don't, chances of us making it are pretty poor, okay? Now the issue is, you cut it so deep, and then the price of gas or something else changes and the drilling activity picks up again—it's how you respond to that.

We don't want to get back into one of these big cycles if we can help it. So we're going to stretch the service industry if the drilling activity is up very much at all in North America. We'll add to it as we go, but we're not going to jump in it with both feet. We've been burned too many times.

I don't believe that you'll see a lot of change by the turn of the century from what we see right now in the mix of dollars spent in the development side of the business versus exploration. Unless here again, they open some of the areas from an exploration standpoint. It will be a very much technology-driven, mature oil business that is lean in structure.

You'll see a lot more partnering between the service companies and the oil companies in order to gain those efficiencies we're talking about. You're going to find much larger independents in North America, and the majors will play a diminishing role because they basically are operating properties they've had for a long time. There's still a big question mark about what happens to the small independent in this country.

1992—a watershed year

While 1986 was the attention getter, 1992 was the key year. I really believe that, because people made a basic decision that we were going to play the cards that we were dealt. We're going to make money in the market that's there, we're going to make money at today's price, and how are we going to do business different than we've done in the past?

That year, in my mind, was probably the key, when we finally got practical and we said to hell with all this other. We're not going to get any help from the government, we're not going to get any help from prices, we're not going to get any help anywhere, we're going to have to do it ourselves. And what can we do that makes some sense, to make money in this business?

We struggled with it all the way down, we struggled with it from January 1, 1982, when the rigs started down from 4,500 or so. We struggled with it for ten years. And we kept thinking it's going to get better. I think, by God, now people say, "If it gets better, that's fine. We'll take that. But we're going to make it as we are today."

ALAN A. BAKER

former chairman of Halliburton Energy Services

Chairman Alan A. Baker's career with Halliburton Company has spanned four decades in oil-field arenas ranging from the Appalachians to Singapore.

A native of Ohio, Baker received a B.S. degree in petroleum engineering from Marietta College in Ohio. Married in his freshman year to his childhood sweetheart, Baker held two jobs and played four sports while studying. Now in his sixties, he still has the easy grace of an athlete.

Baker has received numerous honors for his work, including induction to the Ohio Oil & Gas Association Hall of Fame in 1994 and an award from Spindletop International for outstanding achievements and contributions to the industry in 1993.

It has not always been fun. The downturn forced the oil-field services giant into a massive restructuring termed Project GO (Global Organization), which caused the layoff of a considerable number of workers and the consolidation of the company's business units into Halliburton Energy Services. For a leader who treated the people who worked for him as though they were family, it was a difficult but necessary job.

In our conversation, Baker discusses the need to restructure and the alliances that have resulted within the industry and within Halliburton itself. It is clear that customers have always come first for Baker, and there are plenty of them—about ten thousand in more than one hundred countries, according to a Halliburton estimate.

"I'm a firm believer that we don't get anywhere without a lot of help from other people, and I sure got a lot of help from people all over the world," Baker says. Although he retired as chairman in April 1995, Baker is still active with Halliburton Company, which is the parent company of Halliburton Energy Services, Brown & Root Inc., and Highlands Insurance Company.

Without a doubt, the key issue in the energy industry today is making a profit. It's been very difficult since the mid '80s to assure a proper return on the investment for stockholders and employees as well. That translates into many other things, such as the proper focus we have on our business, what our core competencies are, the proper number of people in the correct jobs. It has many ramifications, but in the end, it all boils down to making a satisfactory profit for the company and a return for the stockholders.

Indeed, the oil and gas industry is the riskiest of businesses. The paradigm of the '70s, which viewed energy prices as ever-increasing, has been replaced with the reality of the '80s and '90s. Talk about risk—well over 500,000 jobs have been lost in the U.S. alone, including very highly specialized people.

Our responsibility is to make certain that the energy industry is a good investment relative to others. The investors are much more astute than they were in the past. They measure returns. They study the focal point of the company. They evaluate the leadership. The directors feel much more responsibility to consider compensation and strategic plans— at just how the company operates.

We must increase the incentive to invest. The next few years are critical, but we have a great opportunity working together—customers, suppliers, and ourselves—to jointly meet the very real issues that challenge our industry today and will determine our future. I'm a believer in alliances and partnerships and teams. We need to integrate the combined efforts of the oil and gas companies with the service sector to obtain the quantum improvement required for industry growth and success. We certainly can help lead the way in what is a very risky business.

There are two primary areas where service and supply companies can help achieve margins beyond that of risk-free investments. These are the costs to explore, develop, and produce, and the quantity of reserves found and produced. Enormous opportunities exist for oil and gas companies and service companies to significantly reduce risk and make major improvements in efficiency by addressing these areas together.

New relationships evolving

There are three key areas of opportunity in which service companies are having a significant impact in reducing risk for oil and gas companies. The first is in new business relationships. Today, our industry is changing the way we are doing business in an effort to influence the service and supply component of the cost per barrel. We are looking at eliminating the inefficiencies and duplications created by the traditional contracting methods. Together we have learned that contracting individual services based on low price has not resulted in lower cost per unit of production over the full life of the well.

Secondly, oil and gas companies and service and supply companies are reengineering. At Halliburton Energy Services, we integrated our ten former business units into a single provider of a broad range of products and services. Now we not only can more clearly identify our costs, but our product and service lines have the incentive to work together to provide cross-functional solutions, rather than fighting for limited resources.

Thirdly, as a single company, we can provide integrated technologies and services at the local level rather that out of a separate company or group. This allows our local managers to customize the relationship according to the operator's local needs, rather than a "cookie-cutter" alliance or partnering relationship that the service company defines. We have teams in place that are working toward technical integration and new business relationships—all with a view toward lowering total life-cycle cost. The focus is on adding value for the customers and for Halliburton.

Service companies are helping to take the lead in this area of improved business relationships. We must go much further together as operators and service companies. In many business relationships, the service and drilling contractors are given specifications and asked to bid to provide the product or service. In some cases, the national oil company acts as an owner, the operating company as a project manager, and the service contractor executes the work. Strictly defined roles such as these can result in duplicated processes, a lack of mutual trust, and a lack of common objectives—all of which increase the total cost of finding and producing hydrocarbons.

We need to continue our search for opportunities for new relationships among all those involved in the development and production of oil and gas. Relationships that eliminate duplication in inspection, take advantage of each other's strengths, and improve both the development and application of technology across the entire group must be explored.

Mutual trust needed

We must develop relationships that are founded on mutual trust and commitment. This allows us to be focused together on reducing the total cost during the life of the field while increasing productivity and ensuring quality. We must also look for new ways to create value in these new business relationships. In some parts of the world, oil and gas companies and service companies are establishing risk/reward scenarios that place a great deal of responsibility on each partner to add value. These arrangements also provide an incentive for successful completion of the project and further rewards for exceeding the goals. Many of these projects have proven very successful for each of the involved parties.

We've seen a number of cases today where the capabilities and information of oil and gas companies are joined with the capabilities of the service, product, and drilling companies in a common pursuit of value goals in a particular project. Many of these relationships are offering major gains in efficiency and cost savings—often producing cost reductions of up to 30 to 40 percent—and they are still improving as the parties gain more trust and the relationships mature.

Another area of significant opportunity for improvement involves the development and application of technology. Today, much of our technology is developed with a great deal of redundancy and at great expense to the industry. Let's say an oil and gas company is spending $10 million to develop the same technology that a service company, another oil and gas company, and a second service company are developing. One may have a slight advantage over the other, but usually only for a very short time. Meanwhile, it has cost the industry $40 million.

Practical use of technology

All of us have believed that proprietary technology provides a competitive advantage. This was certainly true in the past, and is still true in some cases today. But we continue to treat all technology as proprietary. Today, however, much of technology has a very short life and does not provide a long-term advantage for any single company in today's environment. There are great advantages and savings for us if we identify the technology that will have very little competitive advantage and work together to develop it for everyone's benefit.

We also have an opportunity to significantly lower technology costs by developing common nonproprietary or basic technology through consortiums. By using independent labs and universities as well as leveraging technology that already exists in other industries, we can save even more money while improving our technical capabilities. For example, we're using CATSCAN equipment now to evaluate core samples. Much of what has been developed in the medical industry for magnetic imaging could be adapted for our use in reservoir imaging and testing. We need to capitalize on the advances in these industries rather than spending resources on "ground up" development of similar technology.

Frankly, technology application within the context of new business relationships provides a much greater opportunity than the development of technology for technology's sake—which often occurs within a vacuum. Technology integration within organizations and between organizations is the key issue. The value is in reduced cycle time, better information, faster delivery to market, and at lower cost.

There is also a lot of good technology still sitting on the shelf that was developed for a specific project or area. We must begin to work together to communicate and transfer best practices throughout our organizations and throughout the world. We need to effectively apply the technology that is already in place.

The third area where we can improve opportunities is in data integration and management. With the advent of high-speed computers, most of the information necessary to manage the reservoir is in digital form. In the past, with everything in hard copy format, data management was very difficult and prevented integration. But today, with everything in digital format and the vast improvements in hardware and software, we have the opportunity to integrate the data for better information and management of the reservoir.

Data management is a key to lowering costs. We all spend a great deal of time and money trying to integrate the service company or oil company data into our own systems. We must link our resources to come up with a standard data model to be more efficient.

A major challenge for service companies will be to come up with a method to manage and deliver data to oil and gas companies in a form that is readily interpreted so they can make quick and effective recommendations. There is little, if any, advantage in our having different data model platforms. We need to develop a common data model, define the databases in which we will store the information, and establish a philosophy for managing that information.

Dual responsibility

Government also has a role in supporting the industry. It has a responsibility to ensure those lands that belong to the federal government are treated properly. It also has a responsibility to oversee our energy needs, and if the federal lands are properly administered and protected, those lands should be allowed to be explored so we're not so energy-dependent upon other parts of the world.

We need to be responsible, and there will always be oversight agencies in the government. But overregulation is dangerous and very costly. Federal regulations play a large role in our planning, and certainly end up playing a major role in the profit because they are costly.

We all want to live in harmony with the environment. It's important that we act on that desire. Halliburton is developing stimulation chemicals that are biodegradable, more compatible, more environmentally safe. There will be an increased effort to have clean production. That means producing less sand and emulsions so we're able to handle producing problems in a much better manner. Everybody needs to work toward these goals.

In our industry, so many companies have the same challenges, whether it be geologic, geographic, economic, or regulatory. There really is a better understanding of relationships today than in the past, especially between the majors and independents. The majors will still be very prominent in the U.S., because they have so many undeveloped properties—whether it be in the deep, onshore production or the deeper, more hazardous exploration and production offshore. You also have more shallow areas that through 3-D seismic will show such things as subsalt. The majors will divest themselves of some nonprofitable or marginal areas. This could be in partnership with independents or perhaps strictly for sale to the independents.

We see the independent providing a good bulk of the oil and gas from the shallower, less hazardous formations. They have always been good explorers and finders of oil and gas. That's been especially true in properties they've purchased from the majors and incorporated into their own inventory.

Some of the improvements for producers will come from the continued development of directional drilling, which has really assisted the industry. Knowing where the location of the drill bit is in relation to our target allows us to drill multiple wells from a single rig at the surface. That's part of the reason we don't have as many rigs running today. We can drill more wells from one location than ever before and be more accurate in reaching the target.

We'll see other major improvements in the understanding of the reservoir. We are gaining a better understanding of what needs to be done to make it more productive—the necessary chemicals, what pressure rates are required for stimulation, where to stimulate, how to perform the stimulation process to realize maximum production.

I've worked in the industry for forty years and I can remember within the first ten years of my employment that we were told the international scene would far outstrip the domestic scene in revenue and profit. That has been a long time coming. There is tremendous potential overseas. The international scene will grow more important as time goes by—China, Russia, South America, West Africa—there is enormous potential. But let's not forget that there are also hundreds of thousands of wells in the U.S. that have a lot of oil and gas locked up that new technology could recover. The U.S. will always be important.

Attracting young people

It doesn't seem young people are enticed into the industry today. In some ways, it's easy to understand. Young people who look at the history of the oil business see all sorts of things from big cars to diamond rings to poverty and bankruptcy. They see wide fluctuation in the rig numbers, well numbers, and profit numbers. We need stability to attract bright, young people.

The oil and gas business in today's world provides a tremendous opportunity for young people. It's our responsibility to sell young people on the need for education that applies itself to the oil and gas industry, because it has a great future. But it's difficult to find people today coming out of school who want to work in the oil and gas business.

Enrollment is down in the universities. We'll be stretched in the future because as companies downsize, they are losing a lot of experience and expertise. We'll arrive at a point in the future where we'll be at a semi-crisis, because we won't have a sufficient number of really well-educated, experienced people in the oil and gas industry.

Except for a few instances that require special expertise, tomorrow's engineer or manager will have to know much more about the business than today's engineer/manager—how to plan, execute, and analyze projects and processes. Future leaders will also have to deal with a much more sophisticated industry. They will need to evaluate shareholder return when considering strategies and business plans. They will need a broader education and global experience.

We have to lead ourselves into the future. There are enormous opportunities for the energy industry when we work together—the oil and gas operating companies, the service companies, the regulatory agencies. We must work together to assure our joint success. We must understand and apply better business relationships, we must effectively develop and apply technology, and we must attract the best and brightest young minds.

If we accomplish these things, we will help assure the viability and prosperity of the oil and gas industry for many decades to come.

PHILIP BURGUIERES

*chairman, president, and CEO
of Weatherford International Inc.*

Philip Burguieres is recognized as one of the world's foremost oil-field service executives. He is chairman, president, and CEO of Weatherford International Inc., a Houston-based diversified service and manufacturing company that provides tubular handling services, fishing and rental tool services, cementation products, and other specialized equipment.

Since joining the company in April 1991, Burguieres has established Weatherford as the world's largest fishing and rental tool company through a continuous series of acquisitions and consolidations.

A native of Louisiana, Burguieres has a degree in mechanical engineering from the University of Southwestern Louisiana and an MBA from the University of Pennsylvania's Wharton School. He worked for nineteen years at Cameron Iron Works, an international oil service company involved in designing and manufacturing blowout preventers, ball valves, and forged products. When the company was sold to Cooper Industries in 1989, he resigned as chairman, president, and CEO.

Burguieres is extremely outgoing, direct, and decisive. On June 26, 1995, he engineered a monumental merger with competing Enterra Corporation in a $540 million stock swap that stunned the industry, creating the world's sixth-largest oil-field service company with some six thousand employees.

Burguieres' own company provides a classic example of how consolidation is making the industry more efficient by reducing costs.

The energy industry is in for a long and fairly sustained period where we have to assume that oil and gas prices are going to be flat. Now they may not be, but I think we have to assume that.

So I guess the key issue facing the industry when you start from the energy companies, and then it comes down to their service companies, is how you make money in that environment. We have already seen in the early 1980s massive layoffs and consolidations in the oil service industry. That started in 1982, 1983. The oil companies didn't really start their consolidations and cost reductions until the late '80s and early '90s. In fact, we're just seeing some of the major oil companies begin in the last year or two to cut overhead and staff and consolidate.

What that's forcing the oil service companies to do is have further reductions and consolidations, and the trick for the next ten years is going to be to provide products and services that allow our customers to do things more efficiently. So it's not just a matter anymore of cutting costs and consolidating, but to provide *new* products and services that can take, for example, 50 percent of the cost out of drilling a well.

We've already seen that in some areas. You're seeing big platforms in the Gulf of Mexico become subsea systems at a cost savings sometimes of 50 percent, and you might be talking about a billion dollars going down to $500 million. So you're seeing technological innovation save the majors hundreds of millions of dollars.

Consolidating for efficiency

I really think we're going to see some massive consolidations between now and the year 2000. I think amongst the major oil companies—and we're certainly not talking about Exxon and Shell, the two biggest, but down below that—you're going to see an enormous consolidation in the industry. A lot of mid-sized companies in the oil and gas side are going to be buying smaller companies, and in the oil service sector you're going to continue to see enormous consolidation. It's the only way you can become more efficient and survive and make money for your shareholders.

This won't necessarily wipe out the smaller companies. They'll be merged into bigger companies. But there is always room for entrepreneurs in the United States, and there will always be small companies popping up here and there to serve specific purposes in the industry. As these small companies get bigger, you'll find them being bought up by bigger companies.

My particular company in the last four years has made twenty-two acquisitions. We've taken tens of millions of dollars in costs out of our company. And, I'd venture to say, we'll probably do the same thing again in the next three or four years. So you're going to continue to see consolidations in this industry, and it's going to have to become much more cost-efficient. Again, the basic premise behind all this is that we have to assume that the prices of our natural resources—in this specific case oil and gas—are going to stay flat. If they don't, great, but that has to be the working assumption.

Vital resource

Energy in any country is a strategic national resource, but we've pretty much ignored that fact in this country. It's kind of interesting. I deal in forty-two countries around the world, and energy in forty-one of those countries is considered probably one of the three or four highest priorities of their governments. That goes from Canada to Australia to Britain to Norway.

Of those forty-two countries, there's one country that energy is probably down number thirty-eight or thirty-nine on the list, and interestingly enough, that country is called the United States of America.

We really don't have an energy policy in the United States. The last time we had some semblance of an energy policy is when we had an energy crisis during the Carter administration, and that's when he formed the Energy Department, which may now be abolished. And it was really a reaction to crisis. That's the way American energy policy has always worked. It's been a reaction to a crisis.

What should we do? There's a lot of real simple things we *should* do. They're things that the other forty-one countries that have energy policies do. Tax incentives to develop your own natural resources. We should worry about the fact that over 50 percent of our oil is imported right now, and by the year 2000, 65 percent or 75 percent of our oil will be imported. That's a real strategic problem for the United States.

There are a lot of issues with regard to energy that the government needs to be working on. We need to open up more offshore acreage for exploration. A very small part of our offshore acreage is open for exploration. Basically, the whole offshore of California is not open anymore for exploration. Enormous amounts of land off Florida aren't open for exploration. ANWR in Alaska is not open for exploration.

So, tax incentives, more offshore acreage open for exploration, and then all the restrictions we put on people—not just the energy industry—in the last ten years, have been very, very burdensome. But, again, you can look around the world and we're the only country that has a policy where we really don't have a policy with regard to our energy industry.

In Australia, for example, you get tax credits for offshore drilling and exploration. The government, in other words, will say if you go out and spend $500 million drilling for oil offshore Australia and you don't find anything, they'll give you a tax credit against your earnings. Which is an enormous advantage for oil companies to drill offshore Australia. And it's working very well.

The British sector of the North Sea has a number of tax incentives for drilling and exploration. But the main thing is the government, all the way to the top end, focuses on energy policy in the United Kingdom. South America—even in what used to be called third world countries, like Argentina and Colombia—have opened up their energy industry with enormous incentives, inviting people in to develop their natural resources.

There's just a plethora of things that are going on around the world to build up their energy industry. It's kind of peculiar that the United States is putting hundreds of millions of dollars into the former Soviet Union, which used to be our arch enemy, to develop their energy industry, and puts nothing into the U.S. to develop our energy industry.

Higher prices await

Government certainly shouldn't control the industry. The government in many ways should just get out of the way and let the industry operate. By the same token, the government has to look at the total picture of what's going on in the world of energy and worry about things like 60 percent of our energy being imported. The ultimate effect on the consumer will be he'll pay more for his energy in the year 2000 than he otherwise would have. No question about it.

The only thing the government should do is incentivize domestic exploration and production, so that there is a surety for natural security purposes that in the year 2000 and the year 2020 we have our own energy supplies and aren't dependent on other countries, particularly Middle East countries, for our energy—for all of our energy. And that could come, possibly in our lifetime.

We saw environmental issues really come to the forefront with the *Exxon Valdez* and other things. We now will have double-hull tankers bringing oil and gas in from the Middle East, which are very expensive, and an enormous number of restrictions on refineries with the Clean Air Act.

Environmental issues, therefore, are going to continue to play an enormous role in our industry. And, in general, that's good. We should, as industry executives, be very concerned about our environment.

Massive problems

In some respects, environmental issues are going to play a bigger role internationally than domestically. As we've gone into the former Soviet Union, we've found that it's a desert over there in terms of what they've done about the environment.

They have nuclear problems, oil and gas problems, there are pipeline leaks, oil spills all over the countryside. So how the U.S. oil and gas industry that operates abroad is going to handle environmental issues overseas is going to become a real critical issue.

It's an incredible paradox. We are being asked to take care of pollution issues in Ecuador, Colombia, the former Soviet Union, China. Very serious problems. The energy industry is being asked to help solve those problems. And in the United States, we aren't given credit for the things we've done. The energy industry really is a very environmentally sound industry. I know thousands of executives in this industry. We're hunters, we're fishermen. We really care about taking care of the environment.

I do feel that in the long term, environmental issues will not threaten the future of the industry, because what we're beginning to get is a balance now. For a number of years, there were a number of societies, what we'll call environmental associations, that had an enormous amount of power. And if that power had continued to multiply, I think it would have, in fact, threatened the actual existence of this industry.

But I think we're beginning to see reason come into that equation. I'm optimistic that we're going to get this thing sorted out in the next three or four years, and the industry will be able to operate effectively—specifically when you talk about environmental concerns. The two really can work hand in hand. If you look at the middle ground, there are 80 percent of the people who are not in the fringe element of either movement, and they can work very well together. There have been a lot of things that have been done recently to prove that.

But, of course, there's always that 5 percent or 10 percent on the fringe that will never work together. There are people who really don't fully understand when they flip a light switch where the electricity comes from—that you have to drill for oil or gas to put it in the power plant so that you have lights and you can drive your cars and you can do all these things. You don't want to go back to a society where we have candles and no leather shoes and those types of things.

I think the broad spectrum of environmentalists and the broad spectrum of the energy industry can work together. The most interesting example I can give is Robert Redford. He's been involved for probably ten years now in environmental activities, mainly in the western part of the United States, and had considered the energy industry executives and the energy industry in total to be sort of the enemy of the environment.

He has had a number of seminars recently in California to bring energy industry and environmentalists together, and they've gone extremely well. So, when you create a forum where people who are reasonable, which Robert Redford is, can talk to each other, suddenly he starts to realize, "Gee, I didn't know that."

There are a number of things about the oil companies and the energy companies in general that he had no awareness of. By the same token, there was some awareness on the other side. There are more and more of these forums being formed around the United States, where industry leaders are talking to each other. And that's why I'm optimistic.

Becoming number one

When I came to Weatherford, it was about a $140 million company primarily operating internationally in the tubular service business, literally putting tubing and casing in oil and gas wells. We're now the largest rental tool and tubular service company in the world, yet we weren't even in the rental tool business a few years ago. It had a very small domestic presence.

So my first job was to increase the domestic presence of the company and also make it a broader-based company. Again, we had to do that to become more efficient and to make more money for our shareholders.

We decided to get into the rental tool business and liked it, so we bought a couple of small rental tool companies in 1991. In 1992 we bought a very large rental tool company, which made us the largest rental tool company in the world. We continued to buy rental tool companies, and as we've done that, we've made them a lot more efficient. In our tubular running business we've done the same thing. We've increased our domestic presence enormously. Our strategy has been basically to grow through acquisitions, because size brings certain efficiencies with it by taking cost out. So far so good.

Note, too, that as we've gotten bigger, we've been able to put more money into research and development and, in fact, the industry is being forced into technological advancement.

There are lots of examples of it and other things being worked on now that will come to pass. Subsea completions, 3-D seismic, coil tubing—instead of using metal tubing you use composite tubing, forms of plastics and resins—even in my business, we package things together that we couldn't years ago.

Improved packaging

That's enormously important. Not necessarily the technology, but just the packaging. Our customers can buy a whole package of cementation equipment, rental tools, tubular service from us and we can offer them tremendous cost savings as we package goods together.

Packaging and technology are both advancing rapidly, and you're going to see a totally different kind of industry after the year 2000 than you might have right now—technologically and every other way. If you would have asked me ten years ago if our technology would be coming from overseas, I'd have said probably 50 percent of it in ten years, but that hasn't happened. I would have been wrong.

I would say 90 percent of the technology now comes from the United States and the other 10 percent comes from the North Sea. I see that continuing, though maybe the North Sea will be 15 percent and the U.S. will be 85 percent, but most of the technology is continuing to be developed right here in the United States.

The single greatest opportunity in terms of potential is the CIS—the former Soviet Union, which is now the Confederation of Independent States. But there are enormous logistical, political, and economic problems to get oil and gas out of there. I don't see that getting sorted out until after the year 2000. There have been billions of dollars put into the former CIS and very little money has come out.

Then you go down the list. China has a big potential and doesn't have close to the infrastructure problems that the CIS has. South America has become really interesting. Colombia and Argentina, in particular, are going to offer some enormous opportunities for the oil and gas industry in the next five or six years, which have not been hotbeds of activity for the last twenty years.

The whole Pacific Rim is going to be, for my company anyway, a boom area for the next six to ten years. That's where most of the people in the world live. Energy usage is growing 4 to 5 percent per year, and that means all the way from Australia up to the Sakhalin Islands. That's going to be an exciting area.

Within there, you've got lots of exploration. North of Indonesia you've got the Natuna field, which is going to be enormous. There's still lots of activity going on in Malaysia, Thailand, Australia.

Again, government policies are less restrictive in other countries. If you can get a return on investment of 20 percent in Angola versus 10 percent in the United States, you're going to drill in Angola rather than the United States.

Continued reliance

We're going to be using fossil fuels as our primary fuel for at least ten to twenty years, absolute minimum. The nuclear industry has basically been destroyed in the United States and I think it will be that way for at least another ten or fifteen years, before there's any hope of resurrecting it. There's really no other form of energy that can compete with fossil fuels—wind power, solar power—the technology to do that on a massive scale is still years away. Lots of the majors have spent hundreds of millions of dollars trying to develop alternate energy sources, but they haven't been able to do it economically.

People ask if the industry's a good investment. Two things, I think: it is a good investment, but it always has been a cyclical industry. And the cycles will be oil and gas prices. But energy is like food, it's something we're always going to need. I think as a long-term investment, this industry is an excellent investment; but as a cyclical investment, it will have its peaks and valleys. Today, we're at the beginning of a mild downturn in the industry.

For the U.S. industry, gas prices will definitely be the driving factor. No question about it. It's interesting that even at today's gas prices, a lot of companies are doing reasonably well. We have to assume this is going to be the state of the business and learn to make money with low gas prices.

The acquisition of capital really hasn't been as big an issue in the last two or three years as it was in the late '80s, when it was enormously important and difficult to get. Capital has been available to this industry, and I don't see that being a problem in the next six, seven years.

You'll still see a lot more downsizing in the future. It's not over for American industry, period. And it's particularly not over for the energy industry. The service sector probably lost 80 percent of its employment from the 1982 peak, and I think we're going to see some more.

Doing more with less

It's been a horrific human toll on the people losing their jobs. But it just meant you had to make some wrenching management decisions about how you run your company. It's unfortunate, but you've got to do more with less people. People have to, on a relative basis, do more with less money. You've still got to keep on the technological forefront, because you have to provide your customers with more. So it's just been tough. But there is not a shortage of people. You hear that all the time, but we certainly haven't had any problem getting good people to come back into this industry when we needed them.

Looking ahead ten years, I see a lot fewer companies, both among the oil and gas companies and the service companies. But they'll be much bigger and much more efficient as a whole different structure takes shape in the industry, involving a lot more technology and computers.

There will also be different kinds of people running these companies. The so-called good old boy in the oil industry is going to almost be gone ten years from now. The guy who worked his way up from the rig floor to the president of the company— you're just not going to see that happen anymore. They'll be people with degrees who studied petroleum engineering but maybe got an MBA and had to go through training programs to learn what to do. You've always got to have people who know what to do on a rig, etcetera. But it will be those kinds of people who will lead. It won't be people who have literally risen from the bottom to the top.

The leadership position will be where it's always been—with the top major oil companies tending to pull the rest of the industry into the next century. They operate internationally, in nearly every country in the world. They're all in Russia right now. They're all in China. Clearly, they'll be in the leadership position. No question about it.

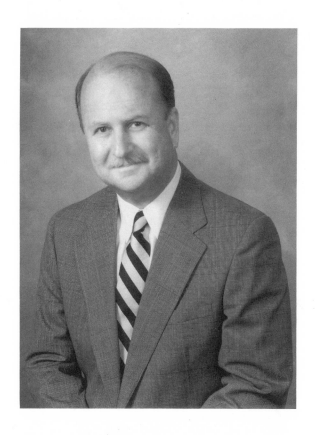

LESTER D. MALLORY JR.

president and CEO of Serengeti International Inc.

Les Mallory is a member of the "new breed." After years of living overseas, he graduated from Texas A&M University at Kingsville and started his career in the natural gas transmission sector of the petroleum industry. He worked in the drilling contracting business before establishing the largest and most dominant marine terminal system along the Texas Gulf Coast.

In 1992, Mallory became president and CEO of Serengeti International Inc., a privately owned Houston oil-field services company that cleans up non-hazardous materials through the process of bioremediation. Today Mallory is recognized as an oil-patch environmentalist who strives to seek the most cost-effective solutions to the industry's needs.

Mallory is also responsible for researching, developing, and introducing white oil technology to the offshore petroleum industry. A strong proponent in getting the "offshore story" told to the general public, he has been one of the driving forces in establishing the Offshore Energy Center, a Galveston-based organization which Mallory serves as vice chairman of the board. He was also a founding member in 1991 of the Texas Bioremediation Council and chairman in 1992 of People for an Energy Policy.

Mallory shows no reluctance in making his views known. He discusses an industry that must learn to work together but that continues to make progress in supplying necessary products while maintaining a healthy environment.

Our overall image and the ability to work together as an industry are extremely critical for us. One must look at the petroleum industry as the largest industry in this country and the world, yet it is the most fragmented industry you will ever find. The unfortunate circumstances are that in its fragmentation, no one entity or person has surfaced in the last thirty years to give it unity and a course of action.

What the industry needs is a unified voice with a great deal of strength behind it. I like to use as an example the television talk shows on Sunday morning—everybody knows about the Archer Daniels Midland Co. and what they have to contribute, but most of the public does not really know who or what the petroleum industry contributes. What the Archer Daniels Midland Co. has done for the soybean, the oil industry forgot to do for gasoline and all of the other products it produces. They're not telling everybody the benefits they brought to them. They're just selling products.

There's a big difference between, "Hey, buy my gasoline," and "Let me tell you how good our gasoline is, why we need it, and how we supply it." Until we start thinking that way, we're still going to be in trouble with our image.

Need for leadership

I find that when it comes to regulatory compliance, the oil industry has not fought hard enough and has allowed itself to be pushed into costly environmental solutions. We are reacting instead of leading. Most of the environmental concerns that come out of our industry, especially the exploration and production sector, are not that dangerous to the environment. They are mainly maintenance problems and should be treated as such.

Understand that the key words are "hazardous" and "nonhazardous." In my opinion, our sector of the industry basically deals in nonhazardous materials. You can clean up diesel fuel, gasoline, jet fuel, and crude oil. These are biodegradable products, reusable products, products you can recycle.

Here's an example. A major refiner in Pasadena had a large amount of hydrocarbon-contaminated soil from thirty to forty years of refinery operation. The soil was saturated. Their first thought—take the soil out and spend a tremendous amount of money to dispose of it in a landfill and bring in new soil. Then came a new approach: Take the soil, recycle it, and use it to build and repair parking lots and roadways through the refinery. That's what they did. They cleaned up their own problem and made a better parking lot by recovering and recycling their own waste products. If people start realizing that you can make a better parking lot, that's when you're going to stop wasting time and money.

Unfortunately, there are not enough of these examples because they are too isolated. And many companies will not share their findings, especially when it comes to environmental solutions. I believe many companies which have environmental problems are not researching cost-effective solutions well enough. They want a quick solution because they want to get some federal or state agency off their back. If it took thirty to forty years to create the problem, then we should be afforded a bit more time to find an economical solution.

We should be going back to the regulators and saying, "Give us six months, or a year or more to do this right." Give one time to answer questions like: "What is it going to do to the local economy? What's it going to do to our company? What is the effect financially on our stockholders?"

Effect of the Valdez

There were two results from that incident, one positive and one negative. The positive is that it made all of us and our industry aware that it had to have a certain set of response mechanisms to take care of cleaning up the spills that are a part of day-to-day business. Whether they are caused by human error or by nature, it's realistic to assume they are going to happen. The incident in Alaska basically said, "Now we know what can happen. We can truly see how big it's going to be and it's impact. Let's learn to protect ourselves and all those concerned."

Small spills can easily be taken care of. We are now all keenly aware of what any type of spill can cause. The oil industry has put in mechanisms and guidelines to protect all of us from these types of catastrophes. And this is what the oil industry has done and done well. That's the positive side.

But that doesn't mean that everyone from dramatic environmentalists, media un-professionals, to the government has the right to hammer the industry. The federal and the state people get a little too carried away. They don't work closely enough.

Overall, I think the *Valdez* spill was the best thing that ever happened, because now everybody is well aware of the daily risks we face in transporting oil. Did it cost a lot of money? Yes. Too much. We all took a beating. And I'm talking about the general public, a tremendous amount of money was wasted. A good example is the present status of Lloyd's of London. The media took liberties beyond professional standards. They created a negative image for an industry which deserves much, much more. And that's the negative aspect.

Remember that these accidents are a result of demand for more and more oil. The more our population grows, the greater the demand. And if we all think the U.S. is a great consumer, just wait until China reaches our level of prosperity. Just give each group or four people in China a car and see what havoc that's going to create. As long as the general public, the media, and government agencies demand their daily fix of gasoline, motor oils, etcetera, we will always have the dangers of transportation. They also must shoulder the blame and responsibility of all our actions.

We need regulation, but not to the point of decimation. We need to abide by a set of clean air and water guidelines. Some things do affect our health on a daily basis, but what I'm trying to say is that we don't have to go overboard on regulations. We can do it in incremental stages. If ingesting DDT and products of this nature are the causes of cancer, then you bet I want to see regulation. It is one thing to diagnose a great many people with cancer and then spend a great deal of time and money trying to find a cure. It's time to spend more of our efforts to find the cause. Toxins coming into our system sooner or later break down our immune system, and we have just so much time before we are in real trouble.

There can be a balanced equation right away between environment and industry, and you don't even have to look at what's regulatory driven. You can have them both and you can have them both economically. It's a matter of attitude.

Most bad environmental problems are abuses—the man rolling the drums down into the river at night—or an accident—that's when a truck driver has a blowout and flips the tanker and spews gasoline all over the freeway. One is basically a criminal act and should receive severe punishment. The other one's an accident and should be treated as such. We're not going to stop delivering gasoline. He had an accident, he flipped it over and put 8,000 gallons on the freeway. Just clean it up. It's no different from spilling milk in your kitchen. Now, yes, there is a difference between milk and gasoline, but you've still got to clean it up. And it's *not that hard.*

Unfortunately, a great deal of the regulations are overreactions by people who really don't understand and by organizations that take in money and immediately go to lobby against our industry. Take this Greenpeace outfit. Twenty percent of what they do is constructive. I want to say perhaps 60 percent of what they do is questionable, the other 20 percent is just plain crazy. This business of trying to stop things happening offshore without having a full realization of the positive benefits that come out of offshore production. These guys can be fanatical and harmful to society in general. Greenpeace is no different in its ignorance than is Representative Frank Riggs of California.[1]

1. Riggs, a Republican, was reported in the *World Energy Update* newsletter as saying that as a member of the majority party, he is in a position to enact legislation to permanently ban leasing of the entire Outer Continental Shelf, including the Gulf of Mexico. He reportedly planned to introduce two bills, the first of which would seek to enact into law an executive order by former President Bush that prohibits new oil and

Evolution of a niche industry

Ten years ago there were no regulations and environmental conscious-ness driving what we do. Ten years ago no one was worried about hy-drocarbon contamination. No one was worried about what stormwater runoff was doing to areas downstream or to the water table.

As populations grow, as water becomes an even more important fac-tor, people start to say, "Wait a minute, we need clean water. Let's change the way we're doing things." In both that change and in new regulations, companies such as ours are conceived and born.

We're in a niche business. We teach you how to keep things clean. Hopefully, we're the type of company that keeps you out of trouble. Many times we get in front of the regulators and express our position on behalf of our clients. Clients need to get a bit more aggressive with the regulator and say, "Hey, you're wrong. We fully understand and can take care of this problem." Many clients don't want to do that. And you're not going to change them overnight.

Given the right treatments, oil will naturally degrade. Nature can and will take care of it, but we need to help nature. We specialize in bioremediation, and therefore we know that by putting the right treat-ment together we can be totally clean in less than ninety days.

We've recently recognized that in all the millions of microbes that exist, there are certain strains of hydrocarbon-eating microbes. These strains, along with specialized nutrient enhancers, have been developed to a state-of-the-art where you can actually say, "I need some for treat-ing diesel fuel," or "I need some for treating crude oil." This technol-ogy is to the point where one can buy these microbes and enhancers to do daily maintenance chores—remember, it's the application and treat-ment that counts.

Now, here's the most important question: Can we get our industry to start applying treatments correctly and put them in the maintenance scheme of things? If we can, you will never see another dirty site in this country. There will never be complaints of another dirty well

gas exploration in California and Florida until the year 2000. The second bill seeks to permanently ban new offshore leasing in the entire OCS, according to the June 12 issue of *Ocean Oil Weekly Report*.

site, drilling site, or production facility if this technology were applied correctly.

Lost in bureaucracy

Sometimes there's too much paperwork, too many demands, too many regulators. You can get up to as many as five regulators on one coastal spill, and you're dealing with federal, state, city, county government. What's worse, they all have a little different twist on the rules.

A prime example is a small operator near Beaumont who had crude oil spill out of his tanks and into a wetlands area. The man knew about our technology and called us. We reacted immediately. The technician we sent ended up having to deal with four different regulatory agencies.

They got a concurrence on how to apply several products to start a bioremediation treatment and take care of the spill. It was approved by the DEQ [the Louisiana regulators], by the Coast Guard because it was near the waters, and by a third group. Everybody was in concurrence, but there was one agency missing, the EPA [Environmental Protection Agency]. They were not on site. After calling the EPA representative in Dallas for his concurrence, his response was, "I don't know a thing about this technology, so the answer is no."

It cost that independent operator ten times the amount to clean up the mess because some EPA underling in Dallas didn't understand the technologies being proposed. So he made a decision to override three other people. Now that small operator could have used that money for payroll, drill some more wells, or increase his production and pay more taxes. But, no, he had to waste it on extra clean-up costs. We all lost on this one person's decision.

Our company's message is to make sure that the industry realizes it is dealing with a company full of ex-oil people, geologists, exploration managers, and drilling people. They understand the industry they came out of after twenty years or more. We think it's important to try to do business with the kind of people that understand our industry, not someone that's just going to give you a high-priced, quick fix and then leave. And that's the problem with some of the environmental companies today.

Bioremediation in some sectors has a very bad name, because people said this "bug" does everything. A salesman can sell a bug that does everything, but if the customer doesn't know how to apply the "bug" and take care of it, it's not going to do anything.

So the salesman leaves after selling several thousand dollars worth of microbes that didn't work. What's your opinion of microbes going to be? They don't work, the technology is no good. When, in reality, the technology is excellent and the most cost-effective. It's how you apply it that counts.

Sometimes we get outstanding responses. Here's a prime example. There's a small operator, Rosewood Resources, which is in the forefront of any oil company that we have dealt with. They are truly first class in their approach to keeping their properties environmentally clean. They called in all their people in a certain area in Oklahoma who were taking care of their production sites. They put them in an office with our people and said, "Let's take care of what we're operating on a day-to-day basis. These gentlemen are here to show us how the products work as well as how to use them. They are also here to train us so the responsibility will be ours from now on as we go to each location. We are also going to be responsible to make sure that the location remains clean."

We cleaned up a small amount of contaminated soil they had, and in the process taught them how to do it. Then they turned around to their people and said, "Okay, now, we're going to do it ourselves." That's progressive thinking. If everybody took that attitude, we would accomplish a lot within five years in this industry. As for attitude and responsibility, Rosewood is at the top of my list and needs to be recognized as an innovative leader.

This has also helped drive technology further. Since we started five years ago, the technology has changed so drastically we can now do certain tasks in half the time. Nutrient enhancers for microbial growth— four years ago, I'd say, was a dormant technology. If you have hydrocarbon contamination, the idea is to get on and off the location as quickly as possible. A microbe can only work so fast. But if you feed it, water it, and aerate it correctly, you can see results a hundred times faster. The nutrient mixed with hydrocarbons which feeds these microbes is the new technology. It has come a long way. We can now really make them multiply.

Technological changes throughout the entire industry in the last three years have been incredible. I think we'll find, on the E&P side, that the cost of producing gas and oil is going to be drastically reduced. We'll probably see no more eyesores the way there used to be. I think you'll find that what we're used to seeing—those large pump jacks that we see from the highway—will be gradually disappearing. There's new downhole equipment being developed. Ours is probably one of the best industries there is when it comes to new technology.

It improves on a daily basis and it can be done anywhere. It can be drilling an offshore well. It can be how we produce one on land. It changes all the time. And that's what makes this industry and this country a leader. This industry helps lead this country on an international basis. I don't care what Washington feels about this industry. They've got it dead wrong. Washington and all its baggage needs to get the lead out and get a good dose of reality when it comes to their views of our industry.

Back to school

One thing that is beginning to turn around is that we're getting more interested students in the industry. We need bright young stars. And they're going to have to come out of our petroleum engineering schools. There's going to be more of a demand for our new technology to be developed with these new students on an international basis. Just look at what China's going to need over the next thirty years. We may go over there and train all of their people, but more importantly it should be 100 percent U.S. products and instructors that are used as the basis for the training.

I think you'll find some domestic companies streamlining themselves a bit more. The North Sea will probably undergo some major changes in late '95 and early '96. The unfortunate part is that certain companies are not retraining their people for alternate needs. They're just closing down whole departments or groups instead of turning around and asking, what can they really do with these people? Can they retrain them and quickly put them on a task force to take care of projects they will eventually give to outside firms?

I feel the industry overall is healthy and on strong footing and in a growth pattern. I believe there's probably going to be more influence from the international arena than ever before—much, much more. The international side is going to become very important to the bottom line to many companies, as far as production, services, and equipment sales. It will put people to work. We're just going to have to deal with a different type of business approach.

I always like to use as an example the Mexicans. They have been proud of their oil fields ever since they nationalized them and threw everybody out. The president of Mexico said recently that under no circumstances was he ever not going to keep the industry whole. It's going to continue to be a Mexican industry.

But isn't it curious how many rigs are doing turnkey work in the Bay of Campeche? And they are American rigs with American crews and American technologies drilling those wells, faster than Pemex ever did. Now, as nationalistic as they are, they had to finally turn around and come this way and use our technology in order to get more production. A large national oil company married to a lethargic and corrupt government does not make for great strides in technology.

At the top

We have great CEOs running individual companies. Take them out of our industry and put them in another arena and you probably will not have the same leadership. I'd prefer to have some of the old wildcatter flavor around who would say, "Get the hell out of the way. I'm going to take care of it." At least that attitude got something done. Today's attitude is so cautious, so filled with legal review.

You are going to have more business types, more financial background types, running the companies, and that has good and bad points. You have to make a profit. You've got to be responsible to your stockholders and investors. The other side is that this type is not going to be as aggressive, and that's my concern. It's going to take some very unique personalities that come out and put an aggressive plan together and say, "Hey, this is what we need to do."

A prime example is what happened to Conoco Oil and their objectives in Iran. What a debacle that turned out to be. Conoco, in my opinion, did the best they could do. The results would have been all to the U.S. benefit. More equipment would have been sold, more people would have been employed, and we would have the most important aspect of all, an exchange between Americans and the "new" Iranian. Any anthropologist worth his salt will tell you that it is from the exchange that we learn from each other. We gain respect for the differences in each other's cultures.

Those farsighted boys in the beltway decided to deny Conoco and the U.S. people the rewards of this business exchange. It ended up benefiting a non-U.S. oil company, Total [of France]. Iran still got what it wanted all along, and we will get nothing to the tune of a probable loss of $600 million or more. How many more farsighted "decisions" will we take from Washington?

Media bias

I think what the media has done is exploit an accident—made it bigger than it should be. Now the *Valdez* was a major accident. But, ever since then, the media has always been on the lookout for any accident. And isn't it strange that when there's an accident, if it's a major oil company which has recognition worldwide, the media attacks it wholeheartedly?

But when it's a small unrecognized company, let's say, transporting oil for someone no one knows, the spill is hardly mentioned and no one is attacked. Now what does that tell you about the media? They're going after the people that have exposure—"logo exposure," that is. Yet when the day is over and they have gotten their piece of flesh, they make no positive statement about the gasoline they are putting in their family car, do they?

I like to call Chevron the one-man band. They are running commercials showing how environmentally concerned they are. They are superb message commercials. They show reefs and sea life and tell you that they're not going to drill there because there's a special coral reef.

They're going to drill some other place. They're promoting themselves as an environmentally conscious company.

On the other hand, Shell Oil is too busy selling its new credit card. Some little dancing credit card that won an award last year for best commercial. Both companies are selling their products, but at least one is also providing a message. We should ask why don't they put together a joint industry message and tell everyone just how correctly we are all treating the environment.

I think major oil has got the backbone and the infrastructure to make it happen and put the leadership together. They have to realize that it's their responsibility. As they become more global, they have even more responsibility. I'd like the majors to sit down and say "We're going to lead instead of react." That we are a premier industry. That we do play on a worldwide basis. That we have the technology that we export every day, and that we're going to let everybody know we are the best.

Untold story

You could tell a story about this industry that's absolutely phenomenal. I'll give you an example. I bet that if you did a presentation on white oil and how white oil is used, you would shock the American public. A good percentage of cosmetics are based on white oil. How many women are putting oil on their faces every day? And where do you think that white oil comes from? It's a hydrocarbon. It comes from the folks who produce hydrocarbons. But I'll bet you when the ladies that we know sit down to apply cosmetics, they don't turn around and say, "You know, this white oil comes from one of Pennzoil's divisions, the same ones who make the motor oil for our car."

Let's talk about grain elevators, just to digress a second. You know that when there's a certain amount of dust in a grain elevator, it becomes a hazard and can literally blow up if there is a spark. To keep that dust factor down, a light spray of white oil is dispensed. Doesn't affect the grain. You and I can continue to eat our Wheaties or whatever else is produced from the grain. White oil has a lot of uses, and one of the biggest is cosmetics.

That story is not being told. At the same time, I keep thinking about that little soybean that I see bounce around on Sunday mornings, and what that soybean does—it makes this, it makes that, it makes hamburgers and God knows how many other good things. What can a barrel of oil do? It can make a hell of a lot more than a soybean.

I'll bet you that if you walked up to every person in this country, he would have something in his pocket made from oil. Whether it's the comb you carry or the plastic shell of a pen. There's something on every person in this country that has a base that comes from oil. But we take it for granted. Everybody takes it for granted.

JOHN M. ELLIS

marketing director of Skaugan PetroTrans Inc.

John M. Ellis has been involved in the shipping sector for twenty-five years and has emerged as a leading spokesperson for crude oil transporters. A graduate of Stephen F. Austin University in Nacogdoches, Texas, Ellis started Gulfco Shipyard in Freeport, Texas, in 1970. He ran that company until it was sold in 1974. Later that year, Ellis became owner and vice president of Canal Service Company, a bulk fuel supplier for the maritime industry.

In 1978, he started Ellis Oil Company, whose overall operations included drilling and lightering. In 1986, he began Petroleum Transfer Limited, which included lightering operations among its worldwide activities. In 1990, Ellis sold the company to Skaugen PetroTrans, whose Houston-based fleet is the world's largest lightering fleet and transfers more than half of all oil lightered on the Gulf Coast. In ship-to-ship lightering, large tankers that cannot berth at U.S. ports are relieved of their cargo at sea by smaller tankers.

Ellis is now director of marketing for Skaugen, which is headquartered in Norway. In this November 1994 conversation, he provides a powerful voice for the shipping sector, which is often maligned by others in the oil industry because of the risks some claim are inherent in the increased flow of imported crude into the United States.

*M*ost oil companies operated their own lightering services as part of their respective marine departments until about 1985. However, due to the increasing specialization, economies of scale, and the introduction of the Oil Pollution Act of 1990, every oil company except Chevron has turned over some or all of these functions to the independent lightering companies.

With the continuing shift toward larger tankers and the ever-increasing need for imported crude to offset the declining U.S. domestic production, the demand for lightering is at an all-time high. With crude imports of over 7 million barrels per day, we estimate that lightering and the Louisiana Offshore Oil Port handle approximately 35 percent of the crude imports, with lightering accounting for 23 percent and LOOP 12 percent of the total imports.

In order to put this into perspective, the 1.6 million barrels per day currently lightered on the Gulf Coast equals the combined daily oil consumption of the Netherlands, Sweden, Norway, Denmark, and Finland. Note that these volumes are delivered using a fleet of only eighteen tankers on a full-time basis, supplemented with additional tonnage chartered in on the spot market. With each lightering tanker making between fifty and seventy-five deliveries into U.S. ports annually, few other companies are exposed to OPA '90 like the three lightering companies.

The lightering industry is genuinely and rightfully proud of its environmental record, which can be favorably compared to that of any port terminal, pipeline, or refinery in the U.S. handling much smaller quantities of crude oil. Since 1982, more than 8,500 lightering operations have been completed without any significant discharge of petroleum. Delivering more than 3.5 billion barrels of crude, the industry has only experienced about a half-dozen minor incidents, each involving a loss of less than fifteen barrels of oil.

The industry's safety record was recently confirmed in an important new research study by Texas A&M University's Professor Von Zharen. She found that neither lightering nor LOOP had experienced any significant tanker accidents over the last ten years, and that lightering has proven itself to be a safe and environmentally sound method for delivering crude oil to the U.S. gulf ports.

Lightering's safety and environmental record is the result of several important factors. The ship-to-ship lightering of oil at sea is a long-established practice with detailed industry-wide requirements for such items as equipment sizes, types, and quantities; the professional staff involved; and the weather window and other operational parameters.

Secondly, lightering is perceived to be riskier than it really is. Loading and discharging through an offshore, ship-to-ship transfer operation is arguably less risky than using a conventional oil load and discharge port. The tankers used in full-time lightering have always been more modern than the world fleet of similar-size tankers. To give you an example, the world fleet's average age of its tankers is 12.3 years, whereas the U.S. lightering fleet is 5.7 years. The world fleet has only 18 percent double-hulled tankers, whereas, in the U.S. lightering fleet it is 39 percent. Double-hulled tankers now account for seven of the eighteen tankers in the trade. Two years ago there was not a single double-hulled tanker in the lightering fleet.

The lightering tankers are generally well maintained and due to the trade are staffed with officers and crews that are very experienced with import maneuvering and cargo handling in general, and in the Texas and Louisiana ports in particular. Offsetting these favorable items is the undisputed fact that all crude is handled twice, and the equally undisputed need for a tanker to actually transit and discharge in an often busy port.

We tend to lose track of the fact that OPA '90 is only a small part of the regulations facing the tanker industry. OPA '90 probably stirred up the tanker industry not because of the complexity or the cost of compliance, but mainly because it so unmercifully outlined the extent of damages and clean-up costs facing a responsible party in an oil spill in U.S. waters. After years of sleeping comfortably in the bliss of ignorance, tanker owners everywhere were awakened and have become painfully aware of the potential liability resulting from negligence.

Let me make my position abundantly clear. Too many shipping companies have been neglecting the quality of their operations for too long. Many tankers were, and unfortunately still are, operated with an overriding goal of minimizing manpower, operating costs, maintenance work, and capital improvements. With flag state controls of tankers virtually disappearing with the advent of flags of convenience, operating costs could be reduced.

With careful management of the classification process, these costs could be even further reduced. And with a typical ship only making money in two or three years out of ten, costs were too often slashed to the bone or even into the bone. We should, therefore, not be surprised when port states led by the U.S. elected to get more involved in determining adequate quality for the tankers trading in their territorial waters.

Here to stay

As a result, OPA '90 is here to stay, and we're already seeing the U.S. Coast Guard allocating more and more resources to its port state control program. As a resident of a coastal community in South Texas, I cannot disagree with these actions. Until we as an industry can bring the quality of the tankers and their operations and the personnel both onboard and onshore up to reasonable and uniformed standards, we will see more and more port states follow the U.S. lead with their versions of OPA '90 and with their own port state quality control programs.

Until we raise the quality of standards to this level, we will not have much input into the many legislative processes to come, especially when the tough issues are being decided. We must face the fact that we as an industry do not have much credibility to leverage on in these matters.

The oil companies have developed a set of minimum standards which they use to approve tankers for use in lightering operations. Even though the lightering companies per se must operate with the same regulatory domain as all the other tanker owners, the industry is unique in that all of its operations are confined to this very limited geographical area.

Contrary to most other ship operators and owners, our company actively supported OPA '90 and even suggested the following: firstly, increasing the quantity of oil containment boom and skimming capacity to be available at each lightering location; secondly, setting a minimum capacity for the emergency pump requirements to ensure that meaningful equipment was available; thirdly, specifying a maximum response time for fire-fighting vessels to arrive on location.

Our reaction to OPA '90 may come as a surprise, but our position was based on the following: OPA '90 would raise the quality of the tanker fleet and the tanker operations worldwide. Newer tankers and more professional tanker operators would lower the probability of a large tanker accident taking place in U.S. waters, which would have a particularly disruptive effect on the lightering industry. OPA '90 would make it more difficult for substandard tankers to trade into the U.S., including participating in the lightering market, which in turn would strengthen earnings. OPA '90, over time, would raise the tanker values and freight rate for modern vessels, thus accelerating the scrapping of older tonnage.

In general, we felt that the tanker industry's reaction to OPA '90 was too emotional and too one-sided. Unlimited liability was not a new concept, and the increased general limits of liability were still comparatively low, even at $1,200 per gross registered ton. In short, we could live with new levels of liability exposure. This does not mean that OPA '90 did not cause us to alter our way of doing business. As an indirect result of OPA '90, the lightering companies have either individually or collectively as an industry modernized the fleet. We've upgraded some lightering support vessels with fire-fighting capability, better hull design, and oil dispersing equipment. We've added important safety equipment, such as secondary fenders and quick-connect cargo hose couplings, to the standard lightering package. Besides having to comply with OPA '90, tougher customer requirements and a more competitive lightering market combined to make these changes happen.

These improvements have come at a surprisingly low cost to the eventual payer of the bill, the American consumer. Lightering rates today are no higher than they were four to five years ago, with lower, underlying tanker rates offsetting the cost of OPA '90 compliance. It has been estimated that the upgraded lightering package has added less than 10 percent to the cost of a lightering, or about 2 to 2.5 cents per barrel for the lightered crude oil. The $1,200 per day Protection and Indemnity [PNI] insurance surcharge for calling U.S. ports and the costs for the new Certificate of Financial Responsibility [COFR, the cost passed on to the tanker industry by the handful of issuing insurance companies] will continue to have a significant impact on our operating margins. Even though a 2- to 2.5-cent per barrel overall increase in lightering costs is insignificant at the consumer level, it adds up to about $10 million per year for the three independent lightering companies.

Additional costs

A portion of this is passed on to customers but, as always, some of the additional cost has to be absorbed by the lightering industry. The need for further fleet renewal, as well as further increases in overall demand for our services, means the lightering industry will be facing large capital requirements over the next five years. As many as ten new lightering tankers will have to be placed into full-time lightering service. A high spec, double-hull, 100,000-ton deadweight newbuilding with desired lightering-specific equipment would now cost $45 million to $50 million each, and may in the next upturn easily cost $60 million each.

We're mandated by OPA '90 that we have to renew the fleet by 2015, but it really kicks in earlier than that because as some ships age, they have to be renewed earlier or they can't trade in the United States.

Additional large investments will be required to renew the lightering support vessel fleet and other related equipment. The industry also faces a large investment in people. The shoreside staffs will have to strengthen to meet the higher oil company standards in areas such as ship management, quality certification, training, and drills. Offshore, we will see improvements as the officers and crews must operate more and more complicated tankers in a very intensive setting program, all within perpetually congested U.S. waters.

Which leads us to the number one question facing the industry at the moment: How is the industry going to finance the $600 million to $700 million in capital investment required before the end of the century, and will the market produce the freight rates necessary to make this happen? The international tanker market is about as close to a perfectly free market as is available today.

The 3,500 oil tankers are owned by some 730 different shipping companies. The barriers to entry are low. All technical, operational, and commercial management functions are available for hire on very reasonable terms. As the supply and demand balance moves in favor of the ship owners, the resulting ordering of new tankers normally limits periods of profitable freight rates to two or three years out of ten. Ship financing is generally short term and not well matched to the life of the underlying, long-term asset.

The financing structure is not designed to support long-term, high-quality tanker operations. As a result, the majority of the profits earned in the oil tanker market is generally by well-timed purchase-and-sell decisions, and too little, if any, is earned from the day-to-day operations of tankers. The industry, therefore, tends to overemphasize the asset-play element and often neglect the day-to-day operations of these ships. This is a big enough concern on the traditional tanker routes, but a real problem in a unique market such as ours. Even though a specialized lightering tanker would be much preferred, all modern tankers effectively compete with the lightering tankers because, in a pinch or if the price is right, they can be used as well.

As the tanker industry adapts to the reality of unlimited liability in the U.S. and the tougher port state requirements everywhere, it will become exceedingly more difficult to survive as a small tanker-fleet owner. The large, well-capitalized shipping companies with sizable owned tanker fleets and large professional organizations will grow at the expense of the smaller companies, leading to a much-needed consolidation in the industry. With the day-to-day operation of tankers becoming more regulated, more risky, and more complicated, the balance

234 / THE OIL MAKERS

between asset play and the operations will shift toward improved day-to-day operations, and the industry will be better off because of it.

The fact that in OPA '90 the owner of the ship is responsible, and in some respects the person who's doing the financing is actually the owner of the ship, means that there are people who will not finance a tanker at this moment until there are some more clarifications or changes.

Since OSHA

I look back at the early '70s, when we were faced with something that industry thought would shut it down, that we would not be able to operate, that the other countries of the world would have an advantage over us, and it would just be devastating to all industries. This was OSHA, the Occupational Safety Hazard Act. But we learned to live with it. It's been proven that we've been better off. The people in the community are better off. The workers are better off, and at not such a large cost to the industry. And if it is a cost, it has probably been beneficial to the individuals out there.

I pray OPA '90 is the same light. And the way I see it, it probably is that way—that it will be a repeat of the scare that we had in OSHA, but in reality it has come around and made things better for us. I can recall seeing tankers that we wouldn't even think of letting come within the 200-mile zone today that were being used just to bring cargoes out of Mexico. I was in Mexico recently and their standards are much higher today due to OPA '90, because they have to trade here. So yes, it was necessary to raise the standards of the tanker industry. But not necessarily the law that was written and passed. I wished that we'd had a little more input into the writing of the law. More thought could have been given to it rather than the knee-jerk reaction. We are going to learn to live with OPA '90. There's no two ways about that.

I was in Connecticut at a marine conference attended by a local congressman, and it was shortly after *Valdez*. He said his staff had come up with the absolute greatest solution that he had ever heard of to make sure that the *Valdez* incident did not reoccur. His idea was to weld springs to the bottom of all ships, so that if they did come in contact with rocks that they would bounce off. My point is, that this is the type of people that wrote our legislation. It's absolutely unfeasible, but he believed this. We see some evidence of this in OPA '90. But again, it's something that's there.

I don't think that it will ever be opened back up, not any time soon. I don't think the environmentalists would like to see it opened back up. The oil companies would not like to see it opened back up. The cargo liability situation, where the cargo owner is also liable for the oil pollution, may in fact someday be that way if we go through the courts. But right now, according to OPA '90, it is the ship owner's responsibility. I doubt the oil companies would want to be responsible for that.

There are very few people that make a profit. And when you have OPA '90 imposing regulations that increase your cost, much of which cannot be passed on to customers, it goes hand in hand. The COFR issue, for an example, is going to cost us somewhere in the neighborhood of $10,000 per lightering. In some instances we can't pass this on to the customer, so it just reduces profits. In the beginning, the first rate schedule said it was going to cost us $50,000 per port call. Competition has come in and it's come down to $10,000. But what we're getting out of this is absolutely zero.

A normal lightering takes three days. We charge about $110,000 to $120,000, depending on the port. We furnish you with a ship, a workboat to bring fenders out to the lightering position, a mooring master, all the bunkers, and so forth. We pay for the port charges. We take you in. The only person you pay is the lightering company. It's kind of a one-stop service.

Small margin

There's very, very little profit. In 1993 we made about $3 million in our operation. But if you take the amount of exposure that we have and the amount of capital that we have invested in our ships, when you pay anywhere between $16,000 to $25,000 a day for a ship [cost to Skaugen], that's not a lot of return for that kind of investment. Multiply twenty-five times three days, that's seventy-five. Your port charges are $20,000, there's $95,000. Then you've got your overhead and your work boat, which is about $3,000 to $4,000 a day. Your margin becomes very slim. Competition's one reason for this. The ones benefiting are the oil companies. Rates are down to where they were four to five years ago.

We recently sent all of our mooring masters to Norway for simulator training at a cost of $150,000. It's very important to continue education, but that's a very expensive item to do in a very small profit-oriented business. We will keep our safety standards up and our quality up. It just becomes harder and harder to do. For a normal tanker owner it's nearly impossible, because they are not making even $3 million a year. They're losing big dollars per year.

I'm a fairly large environmentalist. I belong to a lot of different organizations myself. I don't really find a whole lot of negativism in environmental regulations. OPA '90, I find some problems with it. The Clean Air Act we're going to have to live with; there's some negatives there. But nothing that we can't overcome and live with.

When the ships that we would use in California enter port, they put into the air from their exhaust, emitting a lot of NOX and SO_2— nitrous oxide and sulfurous oxide. They're going to charge you out there for what you put into the air. It could be as much as $100,000 per port call. I think it's going to teach the tanker industry to get scrubbers on their exhaust. And if it's in California, it's going to be here eventually. We may as well get ready for it. I don't know when, maybe after the turn of the century, but it will come.

We are definitely going to have to do additional drilling in the United States. We're going to have to make tax incentives for people to drill. There's a lot of oil in deep water in the United States that's not worth going after, because the price of oil doesn't give you enough return to make you go out there to invest that billion or $2 billion for production. If that's not done, and if ANWR is not implemented, they can't drill in the Alaska North Slope in the refuge. I don't totally agree with that. The oil industry has proven that it's responsible for the environment, and I think that they should have that opportunity to limit imports to us.

If all the aforesaid are not done, we will be importing much, much more oil. With that, it's going to cause a need for not ten new tankers but a lot more than ten new tankers to be out there. That's a tremendous investment for all of us. Costs are going to go up. The level of rates that we see today are an artificially fixed rate for lightering. It will change, and it will change dramatically, as time goes on. In turn, it will start affecting the American consumer. I'm sure that's going to happen. It won't be in '95 or '96. But we're going to start seeing some drastic changes in '97, '98, '99, and the American consumer must get ready to pay for it.

I look back at the Gulf War, when Kuwait was invaded by Iraq. The cost of gasoline in some parts of the country went up 10 cents a gallon and the American public wanted a congressional investigation. These 10-cent raises are not going to be uncommon in the future. If there's a small disruption of the oil flow out of Saudi Arabia, we would be in deep trouble. I realize we have a long-time supply here for maybe 200 days. But still, that would be a tremendous blow and we would all be paying for it as consumers.

As a tanker owner, I should look forward to the additional business. If my business stayed steady like it is today until I'm ready to retire, that would be okay. I need for the rates to go up a little bit. But as an American citizen, I definitely would love to see us not be as dependent on foreign crudes. However, we definitely are and getting more so every day.

I really don't know any tanker owners that I deal with that trade here in the United States and are not as sensitive as we are. To some extent, it's always been there even before OPA '90. I realize that some of them weren't. We've really gotten rid of those people in the United States due to OPA '90.

False image

Most people are environmentally sensitive today. I know that we're portrayed as not being that way. And I see it. I saw the HBO movie about the *Valdez*. I know these men that they were portraying, and that was basically not a true depiction of what went on in the crisis. But again, that's a movie. Most people are very sensitive to the environment. People I know are sensitive to the environment. We've gotten rid of the people that were not sensitive to it.

One of the problems that the tanker industry has is that we have not educated the public on where we are and what we're doing. We [Skaugen PetroTrans] have transferred 2 billion barrels of crude oil and lost only 40 barrels to the environment. This is a record. I can guarantee you can't go to a terminal in Baytown or Texas City that can surpass that record. Nobody can. We're real proud of it. Just equate it to you in a manner that's easy to see for the laymen. If you took the distance from New York City to California and did that in inches or miles, the distance compared to the amount of oil spilled, you would only go 3.5 inches that distance for that 40 barrels versus the 2 billion barrels. That's not very far.

But the public needs to be better informed of who we are and what we are and how we do things. That's been a big problem. A lot of the oil company executives are not familiar with what tankers are and what our safety record is.

We occasionally have an accident, and we've had two that I can quickly respond to. The *Mega Borg*, which was a tanker incident offshore—we were actually lightering that ship when the ship exploded. But it was not a lightering accident. It was a pump-room accident. The second accident was one that occurred in the Houston Ship Channel when a ship ran over a barge. It was the barge that spilled the oil, but it was caused by a ship.

I went to it [*Mega Borg*] the night it occurred. I had to leave the country to go to Egypt. I saw this in Egypt. It was one of the things that was on TV constantly. In my opinion, one reason is that the media look at these because it's attention-getting. The American public has been trained that these are horrible accidents, and they are horrible. But they are very few and far between.

Even in the oil industry, I have heard the same comments—that they're "floating time bombs." But they really are not when you look at the safety record. The safety record of the refineries compared to that of the tanker industry? My goodness, I don't think anybody would really want to compare it. It would make an amazing comparison. Our loss of life in the tanker industry is nil. We did lose four people on the *Mega Borg*, but you don't have an explosion in the Houston Ship Channel that kills twenty-five or thirty people.

These things just don't, knock on wood, happen in our industry. It's extremely rare. I do know that there was a gasoline ship that exploded and, I believe, killed one person in the Houston-Galveston Bay anchorage. But these things are rare. When they do occur, you certainly hear a lot about it. I was reading a little piece in the paper the other day in Austin—a guy died in a tank in a refinery. He was cleaning the refinery tank or cleaning something that cleaned the tank, and he died from the fumes. That's not very newsworthy anymore. But if he'd died on a tanker in an explosion, it would have made national news, whereas this does not. But a life is a life.

The VLCC—very large crude carriers—and ULCC—ultra large crude carriers—have not been operating at a profit for many years. I don't understand how they're even in business, except they do an asset play occasionally and buy or sell a VLCC, depending on the market; make a dollar or two; and then they operate it for a while at a loss. But most of those tankers have been operating at losses. Saudi Petroleum International [the operating arm for ARAMCO for Star Enterprises], which is one of our major importers of crude oil, is very concerned about the ships that they take. They will not take a ship that even comes close. They inspect them, look at them, and still pay them a fairly good price. They don't pay, necessarily, just the market price that's on the spot market. They pay significant numbers for a better-quality ship.

They follow OPA '90 regulations just like we do. Before they can discharge in the U.S. waters, even though they are way out there, they have to have a tank vessel exam letter from the Coast Guard. The Coast Guard issues it for a year, and then they go out and do an inspection. If it's not okay, it cannot discharge into my tankers at sea, even though we're out there.

Cost of Valdez

The *Valdez* incident cost Exxon a large amount of money, and even if the $5 billion cost is there, the tanker industry in itself is paying many, many times over that cost. It has had a very devastating effect financially on the industry, and most of these costs cannot be passed on to the U.S. consumer.

To put the *Valdez* in perspective, I think it was extremely overblown.

I'm extremely sorry for some of the things that happened up there, especially to the environment, but Mother Nature's a very resilient person and she will take care of herself—and this was not done intentionally. Captain [Joe] Hazelwood was accused wrongly, I think, but of course I'm biased toward these people. I work with these captains all the time, and I know it's extremely rare for someone to be negligent on the bridge. Again, it was a tremendous reaction. We had months and months of news on it that, in turn, influenced our Congress. We had knee-jerk reactions and OPA '90 was born.

It has had an effect on us. I wish it had never occurred. We *all* wish it had never occurred. And that was one incident out of how many billions of gallons of oil that's imported into the United States? It's just one incident. Bad as it was—and it was bad in that respect—but I believe it was overblown. I know that some of the people—I belong to Greenpeace—wouldn't agree with me. In some respects, you have to look at the *Valdez*, and what impact did it really have? How do we correct it? And we have taken steps to correct it and not let it happen again.

Let's hope there's not another one.

Other Services

MATTHEW R. SIMMONS

president of Simmons & Company International

There are few deals of any significance in the continually changing oil service and equipment industry that do not go through the offices of Simmons & Company International on the fiftieth floor of 700 Louisiana Street in downtown Houston. Simmons & Company is a widely respected investment bank that concentrates on providing corporate finance expertise to the worldwide service sector. Since its foundation, it has assisted in the completion of about three hundred mergers, acquisitions, and other corporate finance transactions with a combined total value in excess of $12 billion.

The company was cofounded after the 1973 oil embargo by Matthew R. Simmons, its president. Simmons is a native of Utah who graduated cum laude from the University of Utah in 1965 with a B.S. in accounting. He received an MBA with distinction from Harvard Business School in 1967. Simmons spent several years providing consulting and investment-banking advice to a number of clients, one of which was a rapidly growing oil service company. That exposure led him to assist other service companies in mergers and private placements, eventually leading to the formation of his own firm, which now has more than forty employees.

Simmons displays a firm grasp of all aspects of the industry, both domestic and international. If the service sector is indeed the infrastructure of the domestic energy industry, then Simmons & Company has been a crucial force in holding that sector together. But it has not been easy, especially since the downturn of the 1980s, and Simmons believes that even more formidable challenges lie ahead.

One of my biggest frustrations is how badly our firm was blindsided by the collapse of the oil industry. We had a front-row seat in this unbelievable collapse as 1981 came to a close and 1982 began. In retrospect, it is probably fortunate that I did not have a clue what was going to happen or that it was going to last for fourteen years. If I had, I would have been so discouraged that I may even have changed our focus from oil service companies to something else.

As we moved on into the '80s, it became more obvious to me as to what had happened. The more obvious it became, the more annoyed I was that no one saw it coming. It would have been easy for almost anyone to realize that the idea of $50 to $100 oil was insupportable; it did not even have a 1 percent chance of happening. The fact that no one saw a change coming has puzzled me ever since.

After the 1982 collapse, I started spending more time reading and trying to understand what was happening in the macro picture of oil and gas, rather than merely relying on what the oil experts were predicting. I became a strong believer that supply and demand is what always makes a market work, and that it was this relationship that killed the goose that laid the golden egg.

It is true in both North American gas and worldwide oil, because those are the two drivers that impact the oil service industry. Supply and demand had already started to diverge. If you had spent an hour or two looking at readily available information, you would have seen that the oil demand curve, which had been growing for about 30 years, had already peaked and started downward.

Had someone asked, "Why is it starting downward?" or "Is it an aberration or a blip?" you would have been able to quickly lay your hands on all the reasons for its occurrence. It was the twenty-one-mile-per-gallon car (versus the norm of say fifteen mpg), the efficient refrigerator, and all the other things that came onstream as an automatic response to a tenfold increase in oil prices.

The supply side was almost as easy to see, because we had just finished a ten-year exploration boom at all-time high levels. Virtually every basin in the world, including in the Lower 48, had new supplies coming onstream. With falling demand and rising supply, and oil prices already having increased tenfold within a decade, the idea that oil prices could double or triple again was absurd. It is such an easy picture to see. Every time I think about it, I wonder why it was invisible at the time.

I would argue, in fact, that today the fundamentals are just as they were in 1980–81 and, yet again, few people are reading the message. The difference is that it is almost exactly the flip side of the coin. The consensus view that no big changes will take place in the next decade—the view held by many major energy economists and oil and gas leaders—is as crazy as $100 oil was in 1981, and for exactly the same reasons—supply and demand.

Turning point

We are rapidly approaching a classic turning point. In an industry that has been consistently cyclical its entire history, it is ironic that people profess that there will be no change for the foreseeable future. This idea suggests that the cycles have gone out of the oil and gas industry, and that it will remain frozen at its all-time low.

This seems, to say the least, unlikely. If you stick to the basics—supply and demand—the picture is very clear and sharp, and the situation is changing rapidly. I suspect that there are fairly high odds that in the near term we will actually have blown through the "crossing of the lines." When the lines of supply and demand cross—with any commodity—you rarely have flat prices. When demand climbs ahead of supply, it usually accelerates temporarily, because you get the perception of shortage.

To break that spiral, you must do a massive replanting of the crop to correct the shortage. I believe that is one of the reasons why the oil and gas industry has always had long cycles. For instance, you can plant wheat twice a year, but planting a new oil and gas field takes far longer, perhaps five to seven years, when judged from searching for the prospect to peak production. It is almost as simple as that. So, where are we headed as a domestic industry? I think we are headed toward paying the price for the destruction of one of America's most important—possibly even the most important—industry. There will be some pain ahead.

Look at 1982 through 1986, which saw essentially the complete disintegration of the industry. By 1986, we had the biggest gap between potential oil supply and its demand. You can go back and fairly accurately say that you could take all of the consumption of the United States and drive it under a bridge and still have excess.

Since then, the lines of supply and demand for both oil and gas have come closer to equilibrium. Demand for both oil and gas started a relentless growth, fueled in part by low prices. At the same time, supplies of both started a very sharp and steady decline. It is easier to talk about this in terms of worldwide oil, because we can measure oil's supply and demand more precisely than North American gas.

The swing between the rise in oil demand over the last decade and the fall in supply from just two countries alone, the U.S. and Russia, which combined account for the majority of the decline, is close to 18 million barrels a day. Is that a big number? Yes. It is like eliminating the oil production of nine and a half states of Texas, or wiping out 70 percent of OPEC, which is a very significant change. How sustainable are those two slides? How sustainable is the energy demand increase that we have seen? Look carefully at where the demand is being generated. Two-thirds comes from developing countries, which even today barely consume much oil or gas, so it is real and sustainable. Take China, for instance, which is upgrading from bicycles to motorcycles and a handful of actual cars. All use oil-derived fuel.

If you examine the demand side and go back to the 1980–81 picture, where we should have seen the twenty-one-mile-per-gallon car (I use that just as a proxy), there was no way we were going to hold back the decline in demand that was sliding into the system. The picture is almost as dramatic, or even more so, today on rising demand. It is going to take a lot to hold it back. If demand is rising, how persistent is the fall in supply? Is it about over? Not even close. As a matter of fact, it is just getting started.

Fewer well completions

I do not think we are about to turn the oil production decline around. To support this contention, just take a look at our oil well completions. In 1992 and 1993, we averaged about 8,600 well completions, which turned out to be the fourth and fifth years since 1900 when we completed less than 10,000 oil wells. The other three years were 1931, 1933, and 1943. How are we doing in 1994? We are 30 percent under the 1992–93 rate for the first nine months. That does not argue for any kind of a turnaround in our oil production collapse. The irony of the big picture is that demand is on an almost irreversible rate upward, while supply is almost on an irreversible gap downward.

According to the numbers published over the last two or three months by the International Energy Agency in Paris, we are going to have about a 2-million-barrel-per-day gap between worldwide demand and worldwide supply—since Iraq is still embargoed—for the first time since the early 1970s. This is such an unbelievably different picture than what we had to live through in the 1980s. It is almost like we went from the nineteenth to the twenty-first century without passing through the twentieth century.

It is a global picture. However, if you want to switch just to the U.S. (I will stay on oil for the moment), one of the things that has happened to our oil system is that we basically doomed production to start declining when we effectively decided no longer to drill domestically for oil. At the time, I do not think anyone really thought this through, even though it meant we were hitching our economic fate to supplies of foreign oil. We must make sure that we have a delivery system that will get oil here.

It really is not a national security worry of whether Saudi reserves are in good and friendly hands; it is a logistical problem. We must make sure we have "replumbed" ourselves from an oil delivery system that was effectively built around the premise that most of our energy would be supplied out of the U.S., to a new plumbing premise that most of our energy is going to come from sources as far as forty to forty-five days away. These logistics are radically different. There is scant evidence that any significant change in the plumbing system occurred between 1979, when we last had several months of logistical crisis, and today.

Interestingly enough, I read in the paper about the Lakehead Pipeline System in Canada, which brings our single biggest amount of foreign crude into the U.S. The Lakehead Pipeline is about to complete a $300 million expansion, which would basically ensure that Phase I would be ready by December [1994]. This would increase the Lakehead system by 70,000 barrels a day. Phase II would then be ready by the end of next summer, and would increase the system capacity by 170,000 barrels. This process has taken approximately two years to complete, and I bet we will lose 170,000 barrels in the next four or five months just from oil depleting, which is nothing. We consume about 17 million barrels each day.

Had we not put the Louisiana Offshore Oil Port[1] into service, which was really the only response to the gas lines of 1979, we would have maxed out on our ability to import about a year and a half ago.

LOOP increased our deliverability by about 1.2 to 1.3 million barrels per day. About 85 or 90 percent of our port capacity is limited to approximately 85,000-ton or smaller tankers. This is why, other than LOOP, the rest get lightered ashore, using a small shuttle tanker to unload large vessels offshore. If we ever have an accident that causes the

---

1. LOOP, the only deepwater port that exists in the United States, is located eighteen miles offshore and thirty-five miles west of New Orleans. It essentially consists of a huge terminal that allows supertankers to hook up and unload their crude for transfer by a submarine pipeline to onshore storage tanks. The disadvantages of deepwater ports are said to be the costs associated with constructing and maintaining the system, as well as limited flexibility in handling different grades of crude oil.

LOOP to be out of commission for forty-five to sixty days, we would have "lights out" in the United States.

Even if our distribution system and import capability were not issues, we would have another potential source of gridlock, our refining system.

Refinery woes

It is interesting that the last refinery built in the U.S. was commissioned in 1972. Over the last five years, as we introduced a lot of the new environmental regulations, parts of our domestic refinery system were shut down as uneconomic in light of the cost to comply. Once again, as when we shut down our drilling, we made the mistake of not replumbing.

In 1994, all of a sudden, we ended up with about 93 percent average refinery utilization. The year before, we had 91 percent, which was the first time we crossed 90 percent in about thirty years. As far as we can determine, the refining utilization statistics start in 1927, and at only two times were they higher than 91 percent—during the middle of World War II and at the height of the Korean conflict.

When you are running a refinery virtually full out to meet demand, you do not have the ability to perform normal turnarounds [maintenance]. I hear industry people say with some regularity, not being able to perform normal turnarounds is not a problem, because the diagnostics have gotten so good that you can hear a problem coming. That seems strange since every month we read of a fire, explosion, or other incident that shuts down a refinery for some period.

We have trapped ourselves into running close to 100 percent capacity because we have shut the rest of the system down. Talk about an economic straitjacket which will restrain growth. Is it really sustainable? No. Is the drilling collapse sustainable? No.

What sort of shape is our pipeline system in? It was miraculous that the pipeline explosion that occurred March 23, 1994 in Edison, New Jersey, did not kill a thousand people. Had it done so, it would have ranked as our Bhopal or Piper Alpha disaster.

The investigation into the New Jersey natural gas pipeline explosion revealed some shocking things. A thirty-six-inch natural gas pipeline running through the most densely populated region of the United

States exploded with a fireball several hundred feet in the air. It took ten maintenance workers over one hour to manually close the valve, which they had to physically turn 781 times—it had not been turned in approximately a decade. That is not a great comment on how modern our energy distribution system is.[2]

The industry overall is suffering from underexpenditure and from a lack of maintenance. What is really ironic is to look at the strange parallels between the two biggest oil systems in the world, which happen to be Russia and the United States, and see how they are both self-destructing. I am part of a group of at least one hundred senior energy people trying to figure out how to help save Russia's oil and gas system so that the country can survive.[3] No similar group is focusing on the equivalent problem here in the U.S., and it is just as real.

Gas lines

Recently a *New York Times* article included a picture of the gas lines that have recently appeared in Moscow. It was so reminiscent of 1979. I looked at it and thought about how much fun my trip to Russia would be. Sadly though, I suspect there is a fifty-fifty chance that we will see that picture again over the next year or two, and it will not be in Moscow—it will be in New York.

When you finally get to supply/demand equilibrium—and we are so close to equilibrium now that it astonishes me that so few people see it—you no longer have the tolerance for any kind of political unrest or logistical problem. "I am sorry, but the tanker is not here yet," or "Gee, we did not plan on that pipeline explosion," or "How inconvenient that we had a refinery fire," or "The next ten days are going to be abnormally cold weather" will not cut it—the public won't be interested in the excuses. At equilibrium, you have no leeway for dislocations.

---

2. On the basis of a survey of major pipeline companies, conducted as part of the study published in *The Potential for Natural Gas in the United States* (December 1992), the National Petroleum Council has estimated that the industry could be faced with an average annual capital investment of $1.7 billion (1991 dollars) in replacement/refurbishment expenses through 2010.

3. 1995 *World Oil Trends* reports oil production in the Former Soviet Union fell to 6.55 million barrels a day, down from 11.63 million barrels in 1989.

I really don't see any move underway in this country to try and revitalize the industry. I am thoroughly convinced that we will not have one until we have had a disruption. Unfortunately, the sooner the disruption happens, the better. The longer time passes without one, the longer it will take to start rebuilding our domestic oil and gas industry.

I actually think that you could say—forget about oil and gas—democracies have a hard time seeing a crisis approaching and conducting preventive maintenance before it hits. Our country is terrific once we have to react to a tragedy. We are terrible at anticipating and then preventing them. I would be very surprised if we get out of this one without some real disruption, because we have destroyed, essentially ripped apart, our domestic oil and gas infrastructure. All with hardly anyone noticing it in the process, much less being aware of it today.

If I were the person in charge of monitoring the U.S. economy—and I am very glad that I'm not—I would consider that our domestic oil and gas industry should be in the middle of my radar screen. Oil and gas is the equivalent for our economy of oxygen in the human body.

I have a phrase that I like to use, "industrial oxygen." When you think about how important oxygen is to the human body, relative to all the other things the body needs, it is the only one that after about three minutes you no longer need. You can go for a week with no water and a couple of years with no vitamins, but only a couple of minutes or so without oxygen.

If you switch over to various commodity shortages and you run out of copper, I suspect we would see some dislocations showing up somewhere, but I cannot figure out where. However, if you go without oil and gas for about fifteen minutes, the lights go out. Like I said, it is the closest equivalent to oxygen. Since it is our industrial oxygen supply, we had better start monitoring it, because it won't be around forever.

The industry can be revitalized, but it is going to be complicated and hard. It will take longer than most people presume. The pressure is going to be on our side of the business, because it will take time for the oil service industry and equipment manufacturers to gear up for a far higher volume of activities than we are seeing now. That's particularly ironic, given all the pain and heartache involved in getting down to working efficiently at these low levels.

Rig count

One of the remarkable things about the oil service industry is that it is downsized to cope and survive at a 750-rig count. The average oil service company equipment manufacturer is finally making a profit again, which is just a phenomenal accomplishment since we have refitted ourselves as adults to wear a child's suit. All of a sudden, we are going to have a wake-up call and it is going to be very hard to move up the scale again. Technically, if the domestic count is above 850 rigs, we will likely run out of qualified manpower in almost every service in the field.

I do not think anyone knows exactly how large a rig count the country needs. As a simplistic guess, I would say about 2,000 rigs on average each year. Observe the fact that for fifty years we averaged 1,800 rigs at work. For the ten-year period from 1976 to 1986, until we had the great collapse from about 2,000 down to 950 and then to 750, we averaged 2,500 rigs at work. At that count, we had basically reversed the fall of domestic production.

One of the things that troubled me so badly in 1985 was that the industry had the perception that if you could "stay alive 'til 85," things would change. With about 2,000 rigs running, we had seemingly developed perpetual motion and were basically keeping ahead of the S-curve—we were not over-finding. I did not understand why everyone thought we were going to go back to 3,000 rigs at work, because we seemed to be, right now, in perfect balance.

It is going to take a long time to bring people back into the industry and train them. There must be some kind of signal of sustainability, or no one will rush back into the oil patch just to get laid off again in a few months. The average wage of a rig floor crew hand is about 40 percent less today than a general construction job worker is paid. So, why would anybody, unless they just have a passionate romance with the oil and gas industry, come to work on those kinds of terms and know that the odds are pretty high that they would be laid off soon?

Let's talk about the good news that has taken place in this incredible collapse and downsizing. The industry has gone through the biggest technology revolution it has ever seen, and a lot of it came as a response to figuring out how to cope with $15 oil by finding ways to drive drilling costs down.

There are a lot of good news stories that have happened. Just the fact that so many companies found a way to survive, particularly on the service and equipment side, is remarkable. Almost all of the sectors of the industry are going to have to find ways to get a lot bigger faster when the wake-up call arrives. This is going to put a terrific strain on their resources. It will also usher in a danger that we will start losing a lot of these great efficiencies.

We are living in a very unreal world of costs today. The costs have come way down, partially through technology, but also because we have not had to replace anything in a decade and a half. The best example I can give is in a land rig and its drill string, which is like the tires of a car. If you have a good car, but the tires are blown out, it is hard to get anything out of it. Land drilling rigs for the last four or five years have been trading hands for $200,000 to $300,000, but used to cost $6 million to $7 million. We are now nearly out of usable drill string. The average cost of replacing drill string is going to be $250,000. That is less than 5 percent of the replacement cost of the rig.

As replacement costs start hitting the industry, drilling costs are going to start rising again, which also means that oil and gas prices will rise, too. In a sense, we really have been living in a false world of oil and gas economics for the last several years. We perceive our gasoline costs as being higher today. However, if you go back and look on an inflation adjusted basis, they are cheaper than at any time since 1919. This is also why I think that at some point in the not-too-distant future, reality is going to come upon us. Shortages will lead to sharply higher prices, and sharply higher prices will lead to a wake-up call.

Russian dilemma

I think both oil and gas companies are mesmerized with the unbelievable reserves that are in Russia. A lot of equipment suppliers say that it is an enormous opportunity. Even if they did not need to rebuild their system, which is going to cost a tremendous amount, just being able to furnish equipment into that marketplace is a huge task. Plus, the fact that it has always been shut off to us makes it even more enticing. This big opportunity has another side to it, though.

It is not clear in my mind that there is a short-term solution to halting Russia's fall in production. They have broken their reservoirs through poor production techniques, they are completely out of money, and they have no infrastructure within their country left to actually produce any of the goods and equipment needed to start rebuilding their oil and gas industry. They also have years ahead of them where demand, which has been declining at a 15 percent compounded rate, is finally on the verge of leveling out.

The West really needs to make sure that Russia saves their oil and gas system, because if they cannot, it will be far-fetched to think that Russia could remain an economic power. So, there is good reason for that. However, I do not think anyone has factored in the possibility that we might have to save ourselves at the same time.

Will we have the capability and equipment to do both? No. Two or three years ago the Iranians decided to purchase six new state-of-the-art land drilling rigs—and international rigs always have a lot more equipment just because they are so far away from a supply store, etcetera. Six rigs back in the late 1980s would have consumed six days of manufacturing for the industry. By 1990, it strained the U.S. drilling equipment industry just to put those six rigs together. As I watched this happen, I realized it is the best proxy period for how painful it will be to start rebuilding our industry.

Keeping up with gas

It would be nice to think that natural gas could be the answer, and in a perfect world, it should be. It is the most efficient energy source we have available. Everyone knows it is environmentally benevolent and our reserves, by all reckoning, are in phenomenal shape. There is one big flaw in the natural gas picture, which is the question of whether there is any way, period, that the industry will be able to keep up with the astonishing growth we have already seen in natural gas demand by the year 2000. Can we possibly cope with an environment where national demand growth starts accelerating as things like the Clean Air Act enforcement era begin?

If you go back to the turning point in 1986, when all of a sudden natural gas started regaining the market share it lost in the previous fifteen years to coal and nuclear, you find that natural gas demand has enjoyed eight years of a 3.2 percent compound growth rate. If that 3.2 percent, eight-to-nine-year growth rate continues until the year 2000, we would need to have found and be producing the equivalent of 90 percent of a new Gulf of Mexico, or effectively, all that Canada now produces.

In addition to that, we will need to replace close to 70 percent of our current daily natural gas production, which will have depleted by then. That is an unbelievable challenge for the industry. The year 2000 sounds a long way off, but it is just barely five years from now. If we were in the middle of a drilling boom like Canada is, I would say yes, I think we can make that. However, with a 750-rig count, it would be unlikely.

Worry about Alaska

Alaska is a tough picture, because if ANWR is permanently banned, it is believed that unless there is some real surprising find—and hardly anything new is being drilled—that Prudhoe Bay will remain on a slow, permanent decline. It is an old mature field.

I have never heard anyone address the fact that the Prudhoe Bay pipeline [TAP] starts at sea level and ends at sea level, but must climb up over the Brook Range. As the Alaskan pipeline sees lower and lower levels of utilization, at what point do you have to shut down the line to let enough volume build so that you can compress it up over a mountain range? There are very few pipelines ever built that actually cross a mountain range. Will the decline in production of Prudhoe Bay result in a double hit because of the transportation problem?

Nuclear power looks as dead as can be. I do not believe there is any new nuclear capacity on the drawing board. It takes approximately ten years to build a new facility, and no one seems to like nuclear for probably all the wrong reasons. If I were a nuclear person, I would be concerned about the question of "when" rather than "will" we have another Chernobyl explosion out of Eastern Europe or the FSU. The PR impact of that happening would probably set nuclear back another twenty years. To me, it is far-fetched to see nuclear really progressing until the height of our next energy crisis.

Coal, just when you look at the environmental aspects, is a dirty fuel. There is just no way around it. I would hate to be in the coal business, because I see all the problems coming from our refinery business, which seem modest compared to coal. That is why I think gas has such a terrific future, if we can just be sure to keep the deliverability with that.

Environmental folly

The environmental movement is beset with good intention, taken to great extremes without a thought for the consequences. As we learn more about the unbelievable ecological catastrophes in the eastern half of the world and the former Soviet Union, we may start to realize the folly of trying to spend $30 billion to $40 billion to scrub out the last micron of pollutant here in the U.S. versus going abroad and putting out a net to catch the most elemental there. After all, we are a global atmosphere, as the Mount Pinatubo eruption showed, and what happens on the other side of the world sooner rather than later impacts the U.S.

In the oil and gas industry in particular, I think the imbalance between the costly new environmental regulations to the benefits they derive is ridiculous. The Employer Trip Reduction provision of the Clean Air Act of 1990 is a perfect example. In eleven mandated greater populated areas—and Houston just happens to be one of them—every company employing over one hundred people must have a new plan in effect basically forcing people to get to work differently.

Reformulated motor gasoline, which sounds like we are going from sludge to champagne, apparently gets about 2 percent better nonemissions in the atmosphere. It is also 2 percent less fuel efficient, so we will probably use 2 percent more. I just look at the costs to get this marginal improvement without any thought that you are tinkering with industrial oxygen.

I think the oil service industry will be the leader in taking the industry into the next century, where the need will be the greatest. The oil service industry is where all the technology has come from. The leading oil and gas companies—which will basically look, in my opinion, like Microsoft does compared to IBM—will probably not only be the majors, but will include the super independents, which are now as lean and efficient as can be. They understand they do not make wells, they only assemble them through the vast armada of the oil service industry. It only takes about one person to do what the majors used to take ten people to do, and do it clumsily. It will be a combination of the super independents and the technology-driven, lean oil service industry.

Technology has advanced further and faster than ever. However, I think it still has a tremendous way to go. At the end of the day, technology has not replaced the need to drill. The only misuse of technology is the misguided impression that it has eliminated the dry hole. We just keep monitoring the U.S. exploration success rate, which has been flat over the last fifteen years. If we eliminated the dry hole, that would start up—you would see it.

Luckily, the basics of drilling are now radically different. You can justify more rigs at work, because you can pinpoint smaller reserves that you could not see before. However, you must still drill for them. Dry holes will remain a fact of life, so we will need to drill more than we need.

JAMES M. KIPP

managing director of First Union Corporation of North Carolina

Veteran banker James M. Kipp knows something about business downturns. Kipp worked with the energy division of First City National Bank of Houston for thirteen years. He was manager of the division at the time of the bank's demise in 1993. The energy division was responsible for a portfolio of credit commitments extended to companies engaged in oil and gas exploration, petroleum service and supply, product marketing, natural gas transmission, and electric power generation. At its peak, the division had credit commitments of approximately $1.8 billion and loan outstandings in excess of $1 billion.

Kipp, who has two degrees from the University of Texas at Austin, including an MBA in finance, is credited with keeping the energy team together after

First City's collapse. He helped convince First Union Corporation of North Carolina to assume responsibility for the office, which it did in March 1993. Kipp was named managing director of the energy group, which specializes in providing financial services to the energy industry with particular focus on the independent producer.

The energy group's office is still located in the First City Bank tower in downtown Houston. Kipp offers a fascinating and genuinely concerned perspective of the industry from the other side of the table. It is an industry that he has obviously come to appreciate on the basis of, not only his work, but the many petroleum associations with which he is active.

*M*y interest in the industry really dates back to the late '70s. As a graduate student seeking employment, I interviewed with a number of the traditional potential employers who were hiring here in the Southwest. I visited with Marathon, Exxon, Shell, all the majors, for a position on the financial side. And, for a variety of reasons, I opted to go a little bit different course and go the financial side. My exposure has been since early 1980, working with primarily independent E&P companies on the financing side. It's provided a unique perspective, because most of our clients have been fairly small E&P companies, or perhaps small E&P companies that have grown into fairly large E&P companies. I guess it's from somewhat of a unique perspective, because it's enabled me to see some of the challenges that those folks have faced.

What I see as the most pressing obstacle confronting the oil and gas industry, either today or in the forseeable future, I think still remains the key issue which has been an obstacle for this industry since the very beginning. And that's the fact it's a very capital-intensive industry, and there is a tremendous need for capital to keep this industry going. And as a provider of capital, I think any capital source—whether it be a bank, a venture capital firm, or an investment house—plays a very integral role in the overall success or failure of the industry as a whole, as well as of individual companies.

Using that as a springboard, and amplifying a little bit as to what I really see as being the overriding challenges, you could trace that back and go a number of different directions. You could say the environmental challenges are a key issue. You could say that the efficiency of the organization and the cutbacks and the consolidations being driven by this search for efficiency are key challenges.

You could say regulation is a key challenge. You could say public awareness—the plight that the oil and gas industry has faced and will continue to face—is a key issue. You could trace it as, will there be a market for the products that are generated by this industry? That's a key issue. I think they all are very important issues.

But again, where the rubber meets the road, in my mind, is whether there is sufficient capital to fund not only the company on a very specific basis, but also, is there sufficient capital to support the industry as a whole? If you look back to the '70s and the early '80s, clearly the oil and gas industry was an investment vehicle for a number of people to direct their investment dollars. It probably was the one that offered the most opportunity and provided the greatest potential return.

Plenty of alternatives

I don't really see that being the case today, because you have so many alternative investments. You can look to the high-tech field. You can look at some of the things that are happening on the health-care side, with the consolidations occurring there and the same attempts that they're making to generate the efficiencies that are being sought on the oil and gas side. The point is that there are a number of alternative investments for people to direct their investable dollars. When you consider the full array of opportunities, given the returns and the commodity risks associated with the energy industry, it's one that has not been that profitable for many years.

Given the fact that investable dollars are being directed to other industry segments, the big challenge for the executives of the very large as well as the very small oil and gas companies is, How do you generate the capital needed to continue to fund your ongoing operations? You're seeing many different ways of achieving this. Technology has become much more prevalent within this industry. You look at what technology has done for the industry as a whole in the way of 3-D seismic and the measurement by drilling techniques that are being employed on the drilling side of the business.

Clearly, all of these are driven by attempts to better utilize the capital that's available to achieve the overall objectives of these companies. I think a lot of what you're seeing as well on the cutbacks, early retirements, and forced layoffs that are taking place within the industry is recognition, in part, of the advancement of technology.

More importantly, if you're not receiving it through the sale of the commodity—that is, product prices are staying stagnant or declining—you have to operate more efficiently. By operating more efficiently, you are forced to address such issues as, Do you have forty-five accountants when you could get by with only sixteen? or Do you have a team of six geologists when really all you need are two? The exercise people are going through in justifying their cost structure is really a way of squeezing more capital out of their infrastructure that can be redeployed in attempting to generate products to enhance the ongoing livelihood of many of these companies.

Squeezing out costs

After you've done so much of that within your own company that you can't squeeze anything else out, then you take the next logical step and say, "Can I combine my staff with another company's? If I combine these two, what efficiencies will be generated? What efficiencies can be generated from combining my interest in one well and company B's interest in the same well? What are the economies associated with having a much larger interest in a single well?" The result may be that you need only one accountant to cover this property versus two accountants, only one production engineer to cover this property. It funnels throughout the system, the entire infrastructure of the company.

You are seeing this as a logical progression. We've seen this already, and I think you'll see even more as we continue through the next several years. What people call reengineering or rationalization is really an attempt, whether it be in the oil and gas industry or any other industry, to squeeze more efficiency out of their organization, which generates more capital. Ultimately, the basic goal is to enhance their corporate survival.

You have also seen this with regard to the financial sources. Although many and varied, consolidations are prevalent within this industry as well. In the case of commercial banks, you've taken an industry that had 14,000 banks in the 1980s and it's dwindled down to less than 10,000 banks. In all likelihood, over the next five to ten years, that number will be below 5,000 banks. These same trends are occurring with the equity sources and other capital sources that are made available to the oil and gas industry.

The challenge for companies to operate more efficiently is being transposed over a much larger spectrum. What you are seeing in the oil and gas industry in the way of reengineering, retooling, and consolidation is the same exercise that's taking place in commercial banking, investment banking, and any number of industries throughout the economy.

Specialty services

What will evolve from this is a talent pool that will be accessible on a contract basis. You'll have more one- and two-man shops providing specialized services to the industry. Because of the attempts to garner efficiencies and reduce cost structures, there is a genuine concern as to the amount of talent and experience we are losing in the industry.

In visiting with a contractor recently, he told me that, in an attempt to get his cost structure in line, it's been necessary to offer early retirement or severance to his more experienced and higher-compensated technicians. Their positions are being assumed by people with much less experience, but a much lower cost structure. They don't have anything near the same level of hands-on experience that we saw ten years ago.

You don't have the experienced people who have seen the problems that arise when a well encounters mechanical problems. Talent that resulted from many years of experience has been lost. The people being displaced are moving to other industries, retiring and dropping out of the labor force completely, or making themselves available on a contract basis.

I personally believe that the efficiencies that everyone claims are being generated, and the capital being made available, is probably short term. Five years, ten years from now, a lot of folks are going to say, "We did generate some savings initially; but the savings generated were more than offset by the experience lost. It was a period where we simply didn't have the same level of technical expertise and it cost us in the long term. We retrenched and downsized our staff, and lost many talented individuals that had been in the industry for twenty-five, thirty, forty years. They're gone. They're lost forever."

With them, they've taken their experience and knowledge. This has been an invaluable resource which was passed on to those who followed them. Today, you learn by the school of hard knocks. You don't learn by having the old experienced hand who'd seen this five, ten times before. He knew how to address it from practical experience and from those that came before him. These people are gone. This is a real tragedy for the industry, because we're relying so much on individuals with such limited experience.

High cost of downsizing

The exploration side of the business is heavily reliant on the expertise of professionals, such as geophysicists, geologists, exploration engineers, and others. That talent pool has been significantly depleted over the last ten to fifteen years, because, due to the uncertainty inherent in the industry, the influx of young people has fallen dramatically and increasingly the industry is becoming very gray.

The number of students applying to the petroleum engineering schools has declined steadily. The same applies to those studying to become geologists. Fewer college graduates are going to work for the major oil companies or the large independents. There comes a time when you have to ask where the next generation of industry talent will come from. I can't tell you this is being addressed. Unfortunately, the focus of many companies is very short term. They are not focusing on the industry five, ten, or fifteen years from now. In many instances, the philosophy is, "I don't care. I'm not going to be here." The short-term perspective is very concerning.

Attempts have been made to substitute technology—the seismic stations and workstations devoted to enhanced seismic interpretation—to replace this lost talent.

Today, you have a person trying to make an interpretation versus a team of individuals devoted to a basin or a specific project. An argument could be made that because of staff downsizing and the focus on acquisitions, we've lost a very valuable group of energy professionals that were instrumental in discovering oil and gas.

What happened to those individuals? Many are trying to do it independently without the capital sources of their previous enployer. They may be generating ideas for larger companies. But the industry as a whole has experienced a tremendous loss, as many of these people were an integral part of the overall exploration activities. They have been pushed to the sideline because of the focus being directed to acquiring reserves.

As an investment banker and a provider of capital, you are always concerned about the depth of management. In the past it was a given. When you dealt with a company of any size, whether it be a major, large independent, or even a moderately sized independent, you knew there was a talent pool imbedded within the company that afforded you comfort.

It's becoming much more critical for providers of capital to assess the technical competence of the company in addition to the financial aspects of a transaction. Do they have the ability to withstand situations that are somewhat uncommon? You're also trying to determine the depth of technology knowledge, experience, and expertise they possess, which sets them apart from their competition.

Clearly, the industry is becoming more technology driven. However, you can have one of the best exploration companies in the country— they can have a tremendous finding success—but if they lack the technical expertise to manage and produce those wells efficiently, it doesn't mean much. There's more to being a successful oil and gas company than just finding oil and gas. Today, you must find and produce it. More correctly, you must find, produce, and market it efficiently.

Because the complexities of the business transcend the company alone, the strategic alliances and partnerships with suppliers and vendors are also important to the providers of capital. This requires an in-depth analysis of these parties. Do they possess the depth of technical expertise and knowledge to be successful?

Acquisition-oriented

I want to think that companies are using their capital to produce oil more efficiently. What we're seeing, however, and have seen for the last five, seven years, is capital being utilized not so much to create new value, but in value changing hands. More focus is being placed on the acquisition of oil and gas reserves, versus capital being deployed to find new reserves.

Divestitures by the majors, coupled with some companies needing to sell to raise cash and reduce debt, provided a very active acquisition market. That market now is best described as one where there is more money chasing transactions than there are meaningful transactions. If you were to assess how much has been discovered in new reserves, you'd probably be surprised at how little has been discovered over the past ten years.

The key issue is economics. How much is it going to cost me to find it and what do I think I can sell it for? If it costs more to find than I think I can earn, or it doesn't generate an acceptable rate of return, why make the investment? I probably should return it to my shareholders via a stock repurchase or increased dividend. This provides a tangible return to my investors today, versus a promise (or hope) that I will be able to generate a higher rate of return sometime in the foreseeable future.

If you compare the amount of capital that has gone into the industry and been applied to purchase preexisting properties, you would probably be amazed at the ratio of new reserves discovered versus capital going into the industry. It might prove frightening, because a large amount of money has gone into the industry. But it's being driven more as a commodity play with the assumption being that prices will eventually go up. Those who acquire reserves at today's low prices and sell those same reserves in the future at higher commodity prices will prosper.

This may eventually prove to be true, but there are many buyers who have been applying this philosophy for the last five, seven years, and unless they bet correctly on commodity price swings, they probably haven't done too well. Commodity prices for oil and gas have not kept up with the overall rate of inflation. If anything, we've lost ground. So, is capital being deployed efficiently? Perhaps. But if the question asked is whether that capital is being applied to generate new reserves, we would probably be pretty disappointed in the results.

A concern as a financier—in any industry, whether it's oil and gas, autos, or real estate—is that value is being created. If no value is being created, there probably is not going to be much reason for that industry, as well as for you as a financial source to that industry, to be around very long. I trust the oil and gas industry is not like the dinosaur, evolving into something that will eventually go away.

I believe it's an industry that will evolve and better match the overall needs of the world's economies. Hopefully, the contribution we make as a capital source will enable it to make that evolution. Clearly, we hope additional value is created by the exploration endeavors and the new technology being applied to this industry.

But I do not want to sound too negative on reserve acquisitions. From the perspective of a capital source or financier to the industry, what you hope to see is that what you are acquiring has greater value than what is being paid. If what is being acquired has greater value to the buyer because he can operate it more efficiently, or if he has a technical competence enabling him to take these same reserves and enhance the overall recovery rate or marketability, then value is being created. Although no new reserves are discovered, there is an enhancement of value.

Given the choice—exploration or acquisition—you can't point to one and say, I'd like this, or I'd rather have that. We focus on the business entity: the people and management. Are they competent and fundamentally strong in creating value? As they create value, the capital source will likely desire to participate in that success.

The middle class

I don't believe the industry is too large, although I believe we will continue to see consolidations. One difference of the oil and gas industry versus the airline, automotive, and steel industries is that the barriers to entry are not enormous. My prediction is that over time you will witness a phenomenon similar to that experienced in banking. Rather than a standard bell-shaped curve, we are moving to a barbell spectrum of companies. There will remain a number of very large companies and many very small companies. But there will not be many in the middle. The same phenomenon will likely occur in a number of industries. Movement is away from the bell-shaped curve to a barbell curve.

The question raised is whether an industry structured in such a manner can be properly represented at the national, state, or local level by an organization or group that really has two very separate and segmented industry groups having differing agendas. Hopefully, there is some middle ground for the industries to work together. What impacts Exxon probably won't impact a small producer with thirty properties. Any attempt to have one organization representing all participants in an industry as diverse as the oil and gas industry is probably doomed to failure.

Peaceful coexistence

I feel the industry is evolving into one of peaceful coexistence. The majors and large independents recognize that there is a role for the very small independents to play in this industry. On one hand, they provide a ready market for the properties that the larger companies can no longer economically manage or really don't want to bother with anymore. They provide an ample source for smaller participants interested in buying properties. The move toward segmentation within the industry has enabled the larger participants to recognize that there is a benefit to be derived from the small companies, even on the exploration side.

There will remain opportunities for these two segments to work together. Technological advances still originate from smaller shops having geographic or geological specializations. But certain capital constraints will continue to exist. For example, you do not see many small companies participating in a subsalt, where in some instances the participants are spending $40 million for a lease. Although there will remain opportunities for cooperation, capital constraints will remain an obstacle. We may not see as many small- to medium-size independents growing into very large independents, like we witnessed over the last several years. There may not be opportunities for that to occur.

You are seeing a gradual change in leadership. While the technical knowledge and expertise of someone who has been in the business for forty years is still of tremendous value, no one can deny that this business is increasingly becoming more commodity driven. Greater focus is being placed on cash flow. No longer are companies focusing on creating asset value. Increasingly, executives are being forced to focus on how much cash flow is being derived from those assets.

We are witnessing this in a number of industry segments. In addition to the oil and gas segment, it is prevalent in the pipeline and transmission business. We also see it with the utilities. There is greater focus on the financial operations and an interest in seeing management have a firm appreciation of the financial aspects of the business. They may have been on the E&P side for a period of time, but they've also had stints in budgeting, economics, financial analysis, perhaps even an assignment in the treasury area.

People with a marketing focus will also have an advantage. Clearly, the industry is changing dramatically, and it is becoming increasingly important for management to focus not only on finding oil and gas. Rather, the focus must be finding oil and gas economically and generating value for the shareholders.

In the next five to ten years, I'd like to see the industry headed in a direction of making more money. I believe there is real opportunity for this industry to take advantage of the technological advances that have occurred in the last five years and build on them. By exploiting these advances and leveraging off them, tremendous value can be created. Management is keenly aware of the tremendous change that has occurred in this industry.

One important change is that the oil and gas industry is no longer a domestic but a global industry. It took some time for people to realize that they could no longer totally control destiny. There are people on the other side of the world producing a commodity very similar to ours. In many instances, their goals and objectives were completely opposite from ours. But I think the realization has finally sunk in: it is a global industry.

Secondly, the days of the true wildcatters are over. No longer can you afford to drill a hunch and create tremendous wealth with limited or no capital. It's become much more of a financial game—driven by the dollars that are available. Capital availability will have a major role in determining how the industry evolves over the next several years.

Another major change is the commodity aspect of oil and gas. It was pretty simple to run your economics when oil was trading at $2 a barrel and natural gas at 10 to 15 cents per Mcf. As prices have increased, it has become much more difficult for participants in this industry to determine what they will be receiving for the volumes produced.

Hedging risk

Hopefully, as the commodity-hedging products that are being made available through the financial markets become more widely accepted, some of the risk and uncertainty associated with the commodity aspects of oil and gas will diminish. It's no different than what farmers have been using for years in hedging the price risk on their crops. For some reason, the oil and gas industry, particularly the independents, view commodity hedging as something difficult to understand or comprehend. Commodity hedging is a financial tool that many independents should use to mitigate risks.

There are tremendous opportunities for technological enhancement. I am convinced it can remain a very viable industry. One that will certainly have fewer players in the next five to ten years. One that will continue to lose tremendous amounts of talent. Unfortunately, many people at middle-management positions are being displaced by downsizing and reengineering. Consolidations will foster continued displacement of talented individuals.

This country, meanwhile, remains very vulnerable to being brought to the knees by those countries that are the major suppliers of petroleum products. We import over 50 percent of our daily crude supply. This need for crude imports will continue to grow. Five years from today we could be importing 70 to 75 percent of our daily petroleum needs. What is the impact of such a level of imports on our balance of payments? I read that our trade imbalance is approximately $9 billion a month. Over half of this is the result of petroleum imports. It is apparent this situation only worsens as our crude imports continue to rise.

The outlook is not bright for the United States and its energy self-sufficiency. Unfortunately, no one seems to be focused on this issue. As long as prices at the pump are acceptable, as long as gas at the burner tip is cheap enough to sustain economic growth and not put undue pressures on other industries, as long as the industry as a whole does as poor a job communicating the problems facing the industry as the oil and gas industry has done, you will see a continuation of this trend.

DENISE A. BODE

president of the Independent Petroleum Association of America

The leading spokesperson for the nation's independent oil and natural gas producers is a woman whose savvy has made her one of the ranking heavyweights in the petroleum industry. In 1991 Denise A. Bode was selected from more than sixty candidates to become president of the Independent Petroleum Association of America. She is the first woman to hold that office in the history of the IPAA, which was founded in 1929.

Although she does not fit the stereotype of the petroleum industry, Bode has proven on more than one occasion that she can be just as tough as any grizzled wildcatter. Her hard work and accomplishments have won the admiration and respect of the people she represents. A native of Bartlesville, Oklahoma—

home of Phillips Petroleum—Bode has a political science degree from the University of Oklahoma and law degrees from George Mason and Georgetown universities.

A tax counsel for former Senator David L. Boren, a Democrat from Oklahoma, and cofounder of her own lobbying firm, Bode typically works twelve- to fourteen-hour days. The payoffs are increasingly evident. She was influential in the enactment of passages in the Energy Policy Act of 1992 that benefited independents, including relief from the alternative minimum tax, which she vows to see completely repealed. Bode also played a key role in the final language of rate-reform provisions in the Federal Energy Regulatory Commission's revision of natural gas pipeline regulations.

Bode's appointment provided IPAA with a shrewd, well-organized lobbyist who has a strong background in energy and tax issues. She knew the inner workings of Washington at a time when the industry was in desperate need of help. The fact that a woman was now the national spokesperson for the male-dominated oil patch also drew attention from curious lawmakers.

Bode is determined to unify the independents into an effective coalition and to convince the nation of the need for a strong domestic oil and gas industry.

There are independents operating in thirty-three states who produce 39 percent of this nation's crude oil and 65 percent of the natural gas. They are not only the key players onshore though they are replacing the major companies in the offshore arena. They are ideally positioned to find and produce the abundant amount of oil and gas that awaits discovery. There is one problem. There is too much government. Too many misguided policies and regulations prevent this industry from producing the energy needed to fuel our nation.

The ability to form capital, the impact of the tax system, and effects of federal and state regulations on domestic producers are key issues for us.

There have been dramatic changes, particularly in the last five to ten years. It's interesting because many of the big companies are consolidating and looking more like independents.

The oil price drop last year [1993] was the last straw for a lot of people. So you've got a lot of movement toward the middle, from the top down, big getting a little smaller, and some of the small companies have gone away.

Opportunities that improve the ability to form and retain capital for reinvestment are extremely important to independents, because they are constantly trying to improve their bottom line by reducing costs. In fact, that's why a lot of the larger companies try to emulate the independents, because they are lean and mean.

On the regulatory side, it's crucial to do everything possible to reduce that specific cost of doing business so that, again, the independents' bottom line is helped.

We can sell as much oil as we can produce here in the United States. And transportation, whether folks like it or not, is going to be driven by gasoline in the United States. That's going to be our principal transportation fuel, and will continue to be critically important from a national security standpoint as well as a basic economy issue.

Concentrating on gas

The gas side is becoming increasingly important. All the trends are pointing that way. It's not dramatic any one year, but it's going up steadily as more and more people are looking at and relying on gas.

We're only two years into this major change in regulation of the gas industry, so we're still going through some growing pains. For example, changes in the way we store gas. In fact, we just started getting accurate storage figures this year. That's a critical factor. The price fluctuations make it difficult. But if we can increase demand for the product, and natural gas can be more and more a principal part of electric's portfolio, then I think we can have a pretty rosy future.

We're much more sophisticated as an industry. All of us are focused on reliability and not scaring people with a price that would put consumers off the product. We're working together as an industry much better than we've ever done before.

We're competing, obviously, with our friends in Canada, but on our Natural Gas Council we talk a lot about what the future holds, and a lot of folks believe that increasingly our domestically produced gas will be exported to Mexico, because they're still not far enough along. They don't even produce gas principally; it's always gas associated with oil.

That's going to be a new marketplace for us. And it's going to be a competitive marketplace. The Canadians are sending in as much gas as they can. But we're going to be prepared to compete as we modernize our industry with gas bulletin boards, for instance.

There will continue to be mergers, consolidations, and acquisitions, but I also see that there is greater interest in going back to drilling for gas, as opposed to buying it on paper. Of course, all that's dependent on the price.

Living with futures

The futures market increases the instability of the industry. One of the critical factors in our industry is the volatility of the price, and anything that increases volatility is a concern. Everybody talks about the darn futures market and NYMEX, but it's a fact of life and you've got to learn to deal with it. They're not going to eliminate it.

We're not as sophisticated, frankly, in the futures market in the oil and gas business as are other resource industries. The agriculture industry has been playing on the futures market for years, and it's learned how to turn it to their advantage. Now the E&P segment of our business is increasingly learning how to use it as a tool.

But they're not there yet. It's not something you learn in school when you're studying to become a geologist or a petroleum engineer. Things are changing over time. It's still a relatively young industry as far as the futures market goes.

Goals related to a sound energy policy to look for are more certainty and a higher price, but not a price so high that it scares off consumers— a high-enough price, however, that encourages investment and investors. That's critical for us.

The only thing that could dramatically increase the use of alternatives [to fossil fuels] is if the price goes up. And I frankly don't see the price of oil going up dramatically. I would love to see it go up steadily, because our product is cheaper than bottled water and that's outrageous. It causes consumption of a natural resource that should be used wisely.

But right now, less than 1 percent of our energy supply comes from alternate energy, including geothermal and solar. Even if there was a dramatic change in price, the infrastructure needed to support that would take a long time to build.

Energy Policies

When oil prices cratered in 1993 and imports shot up, IPAA demanded the Clinton Administration conduct a national security investigation, which is authorized by Section 232 of the Trade Expansion Act. The Department of Commerce conducted the investigation in 1994, reporting their conclusion to the president on December 30, 1994—that there was a threat to national security posed by cheap oil imports.

Here's the problem: instead of proposing to do something about it, the administration failed to propose any new initiatives that would give the industry the ability to produce more oil and gas resources here at home while reducing our growing reliance on imported oil. The president's inaction is inexcusable.

If left in place, current U.S. energy policies will only further drain our country of its economic wealth. Furthermore, although it isn't politically correct to talk about the national security of America, the report states:

Political and economic problems in the Persian Gulf region make supply disruptions a possibility over the near-term—the United States and the OECD (Organization for Economic Cooperation and Development) countries have limited prospects to offset a major oil supply disruption because 1) there is little surplus production outside the Persian Gulf; 2) U.S. and OECD government oil stocks today provide less protection from an interruption than was the case in 1988; and, 3) there is no substitute for liquefied transportation fuels.

Our tax, environmental, and public lands laws have driven up the cost of doing business in America, making the U.S. oil and gas industry noncompetitive.

We are vigorously advocating the passage of a package of tax and regulatory initiatives put together with the energy industry's support. These include

—a tax credit for old and new oil and natural gas based on production, which is creditable against the regular or alternative minimum tax and is transferable;

—tax credits for existing marginal wells and production from new wells;

—expensing of geological and geophysical costs, of major importance to independents who are putting new technologies to use in the field;

—percentage depletion reforms;

—a new "Royalty Fairness Act," which would improve our ability to do business by imposing a tight statute of limitations on the collection of oil and gas royalties on federal lands;

—lifting the ban on export of Alaskan crude; this would stop the flooding of the California marketplace, which has driven down the price of their oil to $10 a barrel and virtually closed down what drilling they are doing onshore in California;

—legislation to require assessment of actual risk on all environmental rules, protection of private property rights, and repeal of any federal rules that are unfunded mandates;

—improving the enhanced oil recovery credit; and

—modifying OPA '90, the law that would hike from $35 million to $150 million the amount of required financial responsibility for operators on "waters of the U.S."

On OPA '90 relief, we're leading a group of producer organizations to amend the act. We propose defining offshore facilities to include only those producing platforms on the Outer Continental Shelf, and keeping the financial responsibility level at the current $35 million.

In our view, the 104th Congress offers the industry opportunities it hasn't seen in a long time. The House Ways and Means Committee passed [March 24, 1995] a comprehensive bill which includes a proposal to phase out the alternative minimum tax between January 1996 and the year 2000. Slated for early repeal is the remaining AMT preference for intangible drilling costs incurred after December 31, 1995.

Encourage business

We need to work diligently to create a climate that encourages companies to do business in the United States, as opposed to discouraging them. That has a lot to do with tax policy on federal lands. It has a lot to do with royalty policy, and I don't just mean the amount of royalty that has to be paid, but also the degree of difficulty involved in doing business with the government and payment of royalties. I'll discuss that more in detail later.

Regulatory policy needs to be shaped not only in the environmental area—and there are a lot of issues raised there—but also in the public lands area—to provide reasonable access to public lands because we've proven ourselves, particularly in the E&P side of the business, to be able to do it right. The technology is there to be able to do it right without creating a great deal of fear about pollution, and the federal government should give us a chance and quit looking at us like we're an industry of thirty years ago.

For example, there are drilling rigs in Alaska—not a drop—anything that hits that rig doesn't hit the ground. And the reclamation that goes on after the rig is pulled is just incredible. There's so much more improved technology now, but people are slow to give us credit for what we are doing to protect the environment. They should focus their efforts on working together to make sure anything in the past is cleaned up, do that, and let us go about the business of exploration and production. It's certainly regulated more than it needs to be.

We're looking at, particularly in the next two years, a tremendous opportunity to create a little bit more fairness in the way that environmental regulation is made. And this is a long-term trend, I think. The federal government is looking seriously at providing an assessment of the actual risk involved when deciding on whether there should be regulation of, say, drilling muds or injection of produced waters.

This idea of assessing the risk is extremely important. There needs to be an opportunity to go back and look at some of these regulations and make sure that they're worth the cost involved. This is a critical area that's going to be changing, and it's going to provide the business community with a bit more certainty that they're not going to be hit. They feel like they've been squeezed to the very end of their rope. Last year IPAA helped found the Environmental Issues Council, a group of resource and industry associations to focus congressional attention on the need for good science and assessment of risk.

We don't want legislation and regulations based on one picture that's supposed to be representative of the entire industry. The *Exxon Valdez* spill in Alaska is an example. Suddenly, we've got all these drilling rigs being regulated for oil spills when there wasn't a spill on an oil rig. It's overkill. We need to sit down with facts and science.

As I said, there is a real opportunity to turn back the clock on misguided policies and regulations that are doing little to protect the environment. We must work as a team with government, because it plays an important role by providing appropriate oversight. We all gain from making sure that the bad apples don't go out and create a situation that's an eyesore and brings harm to everybody.

Dealing with Interior

One of the major problems is that you've got a huge bureaucracy many times, particularly in the Interior Department. They don't fully understand our business, and yet they spend a tremendous amount of time coming up with regulations that discourage us from doing what they desperately need, which is drilling for oil and gas so they can get the royalties for the federal government.

Every other country spends all their time focused on increasing revenues for the federal government and providing incentives to oil and gas companies to come drill there. It seems silly that they spend all their time trying to prevent us from doing what is in their vested interest.

One of the big problems is that gas contracts are often unknown until a month or so after—an equivalent period of time for oil—so people are paying royalties on an amount they don't know from month to month what it's going to be. Many people have had to set up trust funds that they keep money in just to give the government.

It's a huge mess. They try to create a situation that is just untenable. We have been trying to work with them, but it's very difficult. We're really hopeful that we can get some legislative changes made, particularly in the royalty area, because we think it's in the government's interest and our interest to produce the gas and oil.

We've got a lawsuit pending against the Department of Interior, because they are requiring us to go back eight to ten years and forcing companies to pull out their records of all the settled take-or-pay contracts that were negotiated on natural gas contracts. A lot of these companies are out of business or have been sold, so you can imagine where these records are.

This industry has been completely gutted in that period of time. And yet they're assessing royalties on all those contracts. Just phantom income. You can imagine just the manpower needed. I don't think the government is going to gain that much, and it, frankly, is going to end up causing the industry a heck of a lot more in costs. There's just an incredible amount of bureaucracy.

We'd like to show that the benefit of making these tax and regulatory changes outweighs the costs of these proposals. Texas, Oklahoma, Louisiana, Mississippi, Kansas, Wyoming, to name a few, all have taken the lead in reducing the tax burden on the industry. In Texas, according to state government estimates, the state has actually increased revenue, not lost it.

What is an ecosystem?

Another issue that's interesting, but one that not many people have looked at yet, is ecosystem management, where they're saying that you can't do something because it might affect the ecosystem.

Nobody has really come up with a definition of what an ecosystem is. Every single agency—Interior, Agriculture, Environmental Protection Agency—has different definitions and maps of what defines an ecosystem. Yet they're looking at putting legislation in place that would say, if this affects an ecosystem then you can't do it.

It is another very broad-brush-type issue that could limit development in the United States. It's something we want to look at even more carefully. When they come out and mention this ecosystem initiative, we want to comment by asking, Exactly what is an ecosystem, and what are you talking about?

Open up the offshore

We would particularly like to see the offshore opened up. There are many potential areas that we can get to. Clearly, the industry as a whole would like to see greater opportunities in Alaska and the North Slope. I think that is something that's going to be looked at more seriously in the next two or three years than it has been.

My sense is that opening ANWR is not necessarily a dead letter at all. And I do think that is something the industry obviously would support, even though small producers could not participate. It's an important issue for the country. That's a big new frontier area. But onshore there are lots of areas that are left to be developed in the Lower 48 and offshore in the United States.

We'd like to have an opportunity, and we will have more of an opportunity, if we can change the royalty rules so they're fairer and clearer. That in and of itself ought to provide better access.

It used to be that the majors acquired a majority of the properties and the independents were way behind. The lines have now crossed, and independents are acquiring more properties in the U.S. than the majors by a long shot. As far as total exploration dollars, somebody told me recently those lines are about to cross, too, that independents as a whole are going to be spending more money than majors are in the United States.

I believe the independents are the future of this industry. There will be a lot of majors that will resemble independents, because they have to have a really lean and mean operation to do business here.

What we're seeing here is how you've got an independent that's developed a wonderful niche; they might not have invented it, but they saw the potential and took that technological or geological niche and ran with it, because they are so focused. They don't try to do everything well. They try to pick a few things and then do them very well. Technology particularly is very friendly to them, but it's also very expensive.

They're looking for opportunities and frontiers everywhere. Most of them don't move overseas, because they have focused projects. Their horizons are a little wider today, but they tend not to stray too far from home. We don't have many independents whose sole focus is overseas, but we do have a lot of independents that will have one or two projects over there.

Obviously, they don't have the kind of staff to deal with foreign governments and get information from the State Department or work with OPEC and some of those other groups. We try to interface for them and provide the kind of staff work that they don't have in-house. This is another service that we need to provide for our primarily domestic producers who are looking elsewhere as long as the rules are much tougher here.

Revolutionary changes

We're really part of a revolution of a major domestic industry. And it's hard sometimes when you're in the midst of such a revolution to step back and notice it. A lot of people outside our industry don't recognize it.

I had lunch with a reporter who was fascinated by all these changes and said, "Has anybody reported on this revolution that's going on in your industry?" I said not really. A lot of people will talk about the price and the impact on the industry. And they'll talk about a lot of other things, but nobody's really sat back—outside the energy area—and said, "Wow! You know, there's a sea of change going on in one of our strategic industries."

It's something the Department of Energy folks have focused on, like Deputy Secretary Bill White and others. And they've focused on part of it—especially the technology. But, clearly, we're a part of something that's really an exciting and tough change, because a lot of it means we're still becoming leaner and meaner.

A way of doing business has dramatically changed: the people who are acquiring the properties; the changes in technology, which have made a huge difference in geological concepts; the fact that we're now drilling more gas wells than we are oil wells, and you're seeing gas become more of a mainstay for this industry in the United States than oil—that's a major change; and that we crossed the line in 1994 on oil imports being over 50 percent.

It's been a very difficult time, too, because you can also list all the people who have lost their jobs in this industry—500,000 over the last decade, 22,000 in the first two years of the Clinton administration. Current U.S. policies have drained nearly a half-trillion dollars of American capital out of this country to buy imported oil, and cost the nation tens of billions of dollars in lost domestic oil resources.

There's just so much that people don't understand because they have such a simplistic view. If there is any one thing that we need to be doing as an industry, it is public education. Two years ago we began a program that we're doing on a small scale at IPAA.

We started a Wildcatters Week in the nation's capital, which highlights the exploration and production of oil and gas. All the members of the industry support it as we try to educate policy makers. But we need to take this small concept and turn it into a broad-based concept, just like the cattlemen have with the beef program they're advertising on TV. We need to be doing the same thing as an oil and gas industry.

It's long overdue and, frankly, each of us has got to devote a little bit to that kind of a project and get it under way because there's nobody that has a worse reputation and nobody that deserves it less, really.

JOHN E. OLSON

first vice president, securities research, of Merrill Lynch

John E. Olson, C.F.A. (Chartered Financial Analyst) is considered one of the nation's leading investment consultants on natural gas. As first vice president of securities research for Merrill Lynch, he has responsibilities for oil and gas producers, pipelines, and distributors.

Olson has a bachelor's degree from the University of Pennsylvania and holds an MBA from the renowned Wharton Graduate School. He has closely followed the petroleum industry for more than a quarter of a century, ever since he "blundered into the business."

Olson has worked for Goldman Sachs & Company, First Boston Corporation, Drexel Burnham Lambert Inc., Rotan-Mosle Inc., and Smith Barney &

Company. He was one of the founders of the National Association of Petroleum Investment Analysts and has been on the Institutional All Star Team for natural gas pipelines, producers, and oil service companies.

While sitting at a huge desk that seemed hopelessly cluttered with paperwork, Olson presented a cogent account of an industry in crisis.

I'm speaking from a capital-raising and profitability point of view, and if there were ever a capital-intensive industry that has been living off its past profits, it's been the energy business. To use the old proposition that if an army travels on its stomach, then the energy industry travels on its capital-raising ability.

The biggest problem that this industry has had for the last ten years has been profitability. There have been revolutions in technology, in logistics, in the market place proper. It's been radically restructured in both oil and gas arenas. But the missing link has been the ability of this industry to consistently make a good profit. And that has been nowhere more apparent than in the producing sector here.

We are living through a time in the 1990s which a semifamous economist by the name of Joseph Schumpeter would describe as "creative destruction." The domestic oil industry as we knew it in the '60s and '70s and '80s is rapidly vanishing. There is a pronounced trend to move capital and people overseas, simply because the opportunities to make money there are much bigger and better than here in the United States.

And, indeed, the domestic oil and gas industry seems to be in a death spiral of some proportion, because the rhetoric here has become somewhat circular. Namely, that if you try and make money here in the United States with the hypercompetitive commodity markets that exist, the result becomes profitless prosperity. Since you can't make money here, you go overseas and take even more risks by starting all over again.

Producers and refiners are selling their product at the margin at loss-leader prices, definitely below replacement cost in many markets. The attitude has become that of battle fatigue. Why should we even bother to try in the states if we can't make any money? Therefore, let's go overseas and find a Kahuna of an oil field in some attractive political climates like Russia, Burma, Algeria. This attitude puts the old domestic profitability at risk, and then compounds it by taking shareholder capital overseas.

Needs to make profit

So this industry has not paid much attention to its basic or core problem—its inability to make money. That inability has been reflected in the stock markets by major corrections in the stock prices of many of these companies. We have any number of oil and gas producers which are now simply working for their banks.

Accordingly, something has to change. This creative destruction can go only so far. You cannot go on selling natural gas, in particular, for well below its replacement costs, well below competing fuels, and without any profit margin, for too much longer. Accordingly, we'll see more destruction and more consolidation until this industry learns to price its products correctly.

Currently, if I were a major oil company, I could market heavy fuel oil for $2.60 per million Btus in New York Harbor. Out of another subsidiary, I am currently selling natural gas into that same market for about $1.85. About 75 cents of so-called economic rents—just a free ride—is being left on the table and is going right downstream to the end user.

The producer is taking all the risk and making none of the return, or precious little. And that problem is increasingly being looked at with jaundiced eyes on Wall Street. The late, great George Brown [of Brown & Root] had his bittersweet experience in the oil patch. He said it well: you're simply trading dollars. And you can do a lot better in the world than trade dollars simply by keeping your money in the bank or a money market fund.

So this is the key problem that faces this industry. It's not technology. It's not logistics. It's not regulation. All of those things can be suffered or tolerated or improved upon. But if you don't have a reasonable way to make money for your investors, you won't be around too long.

Corporate America makes, on average, 18 percent return on its equity today. Companies like Coca-Cola make 52 percent on its equity. Gillette, General Mills make 40 percent on their equity. Wrigley Chewing Gum makes 35 percent with almost no debt.

Miniscule returns

The energy industry's oil/gas refining sector generally makes around 5 percent returns on equity, with much more capital intensity than corporate America. Producers are doing much worse. As a consequence, managements have to look at how they are pricing their products. In the 1960s, producers were shutting in production voluntarily or mandatorily because of the old proration schemes. They had a lot of assets on the sidelines that could provide for future growth with a better price structure. That was a mandated market discipline.

Today, there is no market discipline. The commodity markets have essentially gone to the dogs. It's a feast or famine cycle. Currently, petrochemical producers are feasting. Pure producers—oil and gas producers—are starving, as are refiners. So, again, it's a function of changing management styles and pricing techniques. This industry clearly could learn a lot from Johnson & Johnson or Procter & Gamble or whomever, and get away from the mass-merchandising hysteria which pervades the commodity markets here.

The biggest problem in the cargo markets has been their unfortunate evolution. Oil and gas markets lurch from boom to bust and back, because no one has apparently learned anything from history. Therefore they repeat themselves.

Oil companies have always been run by engineers, who have thought that they needed to sell the very last barrel they could produce in order to gain market share. The resulting bureaucracies that have accrued during the post-war era have accordingly been volume-driven. They're paid on volume terms.

Even more bureaucratic

This has been accentuated by the arrival of the futures markets, which have created additional trading liquidity and hedging ability. And the combination of the additional bureaucracies—the trading liquidity and the volume-driven philosophies—have resulted in this unhappy state of affairs. It's a man-made problem. Mind you, if you looked at specialty chemicals or consumer goods and learned your marketing from them, you wouldn't see this kind of marketing chaos repeat itself.

I think that the oil industry will decline rather steadily in the U.S. until it restores the marketing and financial disciplines behind profitability. Old age has set in, and they are dismembering the assets piece by piece. Balkanizing it. The major oil companies are getting out of the producing side of business. Refining has become a closely regulated kind of business, but without any financial protections. It's almost gotten to the point that you can't go to the refinery bathroom without a permit now.

And the disincentives to reinvest money in the Lower 48 are probably greater than the incentives now. It's a snowball effect. The difficulty in making the necessary profits becomes exponentially greater when you pile on regulatory burdens, reporting, and legal burdens. We may have come to the point that it has made no economic sense to build any new grassroots project or major industry project in the United States.

With little prospects of seeing a refinery getting built anytime soon, you may see more petrochemical and refining debottleneckings and the like. But they are pale in comparison to what we saw ten or twenty years ago. It'll be a fraction of that. I think this industry is becoming a collection of dwarfs and orphans.

Increased fragmentation

It may even come to pass that this industry steadily restructures itself into a cottage industry. That is probably much more visible on the exploration and production front today than it is in some of the other sectors. But within five years' time, we may be looking at a much more fragmented industry. And hopefully a healthier industry as a result. The more you fragment the industry, the greater the need to have a good marketing discipline to make money to cover your interest expenses and debt service that way. In contrast, the major oil companies have traditionally subsidized their weaker links over time.

You may have more regional refiners, regional marketers, people who operate in Texas and the Southwest or the West Coast, for instance. I don't see this industry as one which can carry the current bureaucracies with such low margins that pervade the 1995 cargo markets. So there will be more medium-sized players than before, and perhaps fewer large competitors. You may see this sort of diaspora, this sort of opening up, in an opportunistic sense that's been under way for the last few years, as this industry continues to reinvent itself and its profitability.

The majors have clearly taken a backseat in the Lower 48 exploration and production arena. They still dominate the refining marketing arena, if by no other reason than default. Nobody else wants the particular assets. Just over Christmas we saw one of these defaults occur when Chevron was trying to sell a big refinery for a very low price to Clark Oil. But Clark could not arrange the financing. And another refinery has come out of bankruptcy, but had to raise money at an interest cost of 16.5 percent in order to fund that startup.

So perhaps the major oil companies will partially rationalize their asset base to only those arenas where they are profitable and have a chance to survive. And they will, very likely, try and parcel out the rest to people who can either use the assets better, or else just write them off. For the moment, I think most of them have a fairly well-defined agenda. The issue is one of the major oil companies restructuring themselves, creating stand-alone subsidiaries or selling to the public like ARCO did last year, with a company called Vastar Resources, on extremely profitable terms, to ARCO. Devices like that seem very plausible to me.

Era of the independent

The independents are the future of the domestic exploration and production in this country, period. They will dominate the onshore market. They will increasingly tiptoe into the offshore markets and, using the advantages of new seismic techniques, they will be able to go where majors have gone before and do better there.

The major oil companies, interestingly, should operate more like independents in order to survive. The independents have the strong backing of Wall Street now, but how long that lasts with this profitless prosperity remains to be seen. It will all be a function of price. If you enter a period of prolonged price weakness, especially for natural gas, it's likely that a scenario of strong consolidation into mega-independents will occur. If the price structure does rebound, you will still have a number of mom and pop shops out there.

I also believe the independents will remain acquisition-oriented. That has been the name of the game for the last five years. Most independent friends of mine in Houston wake up in the morning figuring out how they are going to acquire another independent.

You are witnessing only a trading of dollars there. Is there any real value added? No. Do bankers, accountants, and lawyers benefit more than anyone else? Yes. The massive asset-swapping and reshuffling that's going on is all being done on the hope that the price of oil and gas respectively will go up over time and exceed both inflation and the cost of money, too. And as long as they are able to arrange capital backing by a two-step transaction, borrowing the money to do the deal and then get that money taken out by equity offerings on Wall Street, independents will persevere.

I would say that it is probably a healthy effect, although there are a lot of people in this business operating on a wing and a prayer. I think that you probably need to have a minimum asset size to be effective in this industry. There are thousands of independents which operate only a small flock of wells. Maybe there is some magic asset size between $5 million to $100 million out there. The small independents have been written off countless times, but they are still there.

Finding capital

Capital is available at a price, and sometimes you have to take a big penalty or take a big haircut to raise that capital. Mostly at the end of the day, it comes from Wall Street. First of all, through bond offerings or credit lines or whatever. But more importantly, it gets it from equity investors. That's the permanent capital of the industry, and that is being rationed very carefully nowadays.

Parts of the industry should remain very good investments because they have developed strategies that work, at least for the near term. And I would imagine that because of the boom-bust cycles, various other parts of the industry will fall into place as the cycles tighten out there.

Investors prefer the niche sectors nowadays. The pipeline industry looks fairly attractive, because its economics have been largely bullet-proofed by the regulators. The natural gas liquids industry—that's about 20 percent to 25 percent of our overall domestic crude and liquids production—has been a solidly profitable business for years. I would say that those clearly have the advantage near term. Down the road, if wellhead prices get better, then we'll see some renewed emphasis in exploration and production.

But that sector [exploration and production] is clearly hamstrung, bedeviled by restrictions on land areas; national forest areas; and places like the petroleum reserve [ANWR] up in Alaska.

Environmental overkill

ANWR is critical, because it sits right next door to one of the largest discoveries in the past fifty years of oil reserves. The ecological damage that would be done to ANWR is miniscule in relation to the economic benefits. This is a classic case of environmental overkill. Classic because this is something no ordinary human being is ever going to visit in their lifetime, and there is virtually nothing there, except oil. Once we stop playing with political symbols and imagery in Washington and get down to the harsh realities of trade deficits and national security, ANWR will perhaps get developed.

I'd say that it's really a whole suite of regulations emanating from the Bureau of Land Management, the Minerals Management Service, in tandem with Clean Air Act amendments and EPA regulations where you have to be purer than the pope in terms of drilling a well.

The whole permitting process now in some areas has gone from the ridiculous to beyond. Whenever you need archaeologists, botanists, biologists, and whatever to tell you if you can drill a well or lay a pipeline, you know to what depths of political correctness we have descended.

That balance between the two sides is yet to be found. But it is to be found more on the front lines than it is by Monday morning quarterbacking at the EPA in Washington, for instance. I dare say that this industry continues to come down a learning curve and, with some healthy prompting from various regulators, has learned how to police itself. But there's still a long way to go out there.

The government should maintain a level playing field in a regulatory sense. And it should figure out ways to repackage its presence in the form of light-handed regulation. There needs to be a synthesis in the permitting process. Instead of simply building a pipeline or drilling a well, you now have to go through a regulatory maze to figure it all out and get the right environmental permits done. That could probably fill up this entire desk to the ceiling.

That has to be probably the greatest burning issue for government at this point, to somehow become more user friendly at both federal and state levels. It is very hard to drill wells in certain parts of the country now without being regulated to death.

I think the mindset will change. Bureaucrats are either forced to change in tandem with their industry, or they are forced to change through legislation. It is too early to see what kind of positive effect the incoming Republican Congress will have, but I think people's hopes should be up somewhat.

A sound energy policy involves a recognition of the three constituents here—the industry, its investors, and its regulators. If this country wants to maintain a viable, domestic producing, refining, and marketing and pipeline industry, then Washington policymakers are going to have to be more statesmenlike about regulation and the profitability thereof. On the other hand, the industry's mentality on a number of issues is almost claustrophobic, because it has been sued so many times. The national energy policy debate has become a classic case of battle fatigue.

Technological advances will continue to be very important. They have moved on the exploration and production front in quantum leaps. Now we're going to see refinements there until the next leaps occur. I'm not well-versed enough to tell you just where that next step will come—I'm not sure anyone is—but I think that there is still enough left on every front to improve, whether you are talking about gasoline efficiencies all the way to fuel-conservation efficiencies and demand management of energy.

The availability of money for researching advanced technology is a growing concern. The research and development powerhouses that were there in the '60s and '70s, for instance, are no longer there. The fine Phillips Petroleum, Amoco, and Exxon research facilities, most of these have been decimated, and so has the money available to fund the more exotic R&D efforts.

Look to conglomerates

I think the so-called energy conglomerates will be leading the industry into the next century. If you look at some of these companies out there, they are into any number of different energy businesses. Tenneco, Coastal, Enron, Nova in Canada. They're in petrochemicals and in refining, natural gas pipelines, coal, exploration/production, and power plants.

These people have a much better understanding of what Wall Street mandates are all about: namely, making money for your investors. And they are driven by profit motives. They are the latter-day profit maximizers. They have learned to make money in the energy business. These are the people who will inherit the Earth out there. The strategy is likely to be more of a conglomerate approach to life, where you can make money in certain niches or businesses and you leave the rest of the scraps to your smaller-potato competitors here.

SAM FLETCHER

former energy writer for the Houston Post

Veteran energy writer Sam Fletcher was born in the small town of Gladewater in the heart of the giant East Texas oil field, the biggest in the Lower 48 and second nationally only to Alaska's Prudhoe Bay. Fletcher's family consisted of oil-patch workers for three generations. "We were oil-field trash back when it was an insult instead of the Texas bumper-sticker brag of the 1980s," he says.

Fletcher spent his summers in jobs with seismic exploration crews, working on the powder (dynamite) truck, pouring seventy-five pounds of nitrate down shot holes, or as a "jug hustler," laying and picking up geophone lines to record those shots. He later worked as a roustabout, doing routine maintenance at well sites and gas processing plants in West Texas. He was also a geological lab assistant for the former Humble Oil Company in Midland.

"But I steered clear of the better-paying roughneck jobs to avoid getting trapped in the oil patch," Fletcher remembers. "Figured I could always walk away from the lower-paying jobs if I ever got a chance to go to college." Which he did, receiving an undergraduate degree and a master's in mass communication from Texas Tech University in Lubbock.

Fletcher joined the *Houston Post* in 1973. Four years later, he was assigned the oil and gas beat, where he won numerous reporting awards. After the *Post*'s demise in April 1995, he began covering the industry for a trade journal.

Fletcher is known throughout the industry and the media for his knowledge, gracious manner, and honesty. He provides a passionate glimpse into an industry that he has seen from many sides.

*T*he energy industry is faced with several key questions for the remainder of the '90s. They are government regulations, availability of capital, and the influence on operations of environmental issues.

Environment is a major buzzword, particularly for the integrated companies—the way it impacts their downstream operations with the reformulated gasoline they've just spent so much money on. They're having to revamp many of their refineries, and there's major spending needed to bring emissions into compliance with federal standards. This means a lot of money that could be used elsewhere is going to be spent on environmental issues.

Something interesting arose during the recent Arthur Andersen energy conference. One day, Al DeCrane, chairman of Texaco, spoke. The next day it was Lawrence Fuller, chairman of Amoco. A good portion of DeCrane's talk and virtually all of Fuller's talk centered on the upcoming federal requirement that wants employers to decrease the number of miles their employees drive by 25 percent.

Of all the things you would think they would be addressing of concern, both of these men, independently—because I doubt they got together beforehand and did this—came up with this particular issue.

And the reason they were talking about it instead of addressing what we can do about cleaning up the air is that it's the government mandating that you will reduce the miles driven by a certain amount. Their argument is, this is part of some nefarious plot among staffers and "tree-huggers," etcetera, to wean people away from autos and fossil fuels to whatever alternatives there are. This is the way they view it.

Their argument is that maybe under the new Congress, people will start taking a closer look at the cost benefits of regulation, which is a good argument. That's something that people ought to be taking a closer look at instead of just going out and mandating.

For instance, they were required to put in special fuel systems at service stations to suck up gasoline fumes. But with new cars coming on-stream along with reformulated gasoline, in a very short time that's going to be a moot point, because there's not going to be many vapors escaping. So why this big investment in gasoline vapor-recovery systems when you already had reformulated gasoline coming in and new cars being required to retain vapors and eliminate them through other systems? It's redundant spending and takes capital out of the industry that could be going elsewhere.

They have a point. But it just struck me as funny that of all the other things going on, that this particular issue—the employee mileage thing—is the thing that captured the attention of two chief executives of two large major companies.

Government regulation in all forms and shapes is going to continue to be a major issue. I know the oil companies, particularly the majors, have complained for so long instead of trying to educate people that it's become sort of a permanent whine. They get written off, "Well, that's the oil industry. Of course they're going to complain." They've got some very good points, though.

Targeted

There is no other industry in this country that has been subjected to as many rifle-shot regulations as those aimed particularly at this industry. You could not apply a windfall profits tax to the steel industry or to the auto industry that was heaped on the petroleum industry, and maintained even after the so-called windfall profits had long disappeared. It was an excise tax that was directed at a single industry. I agree with a lot of people in the industry that it was really unconstitutional. They did challenge it on the grounds of constitutionality and lost.

No other industry has to break down their overseas earnings to the degree that the oil industry does. How much did you earn on exploration and production? How much on refining? How much on marketing? That's what the majors have to do.

The result, in effect, is double taxation oftentimes. No other industry in America has to do that. Yes, they are overregulated. It's the biggest industry. There's more money involved in that industry than any other, and it's seen by too many people in government, particularly at the federal level, as being a cash cow.

I agree with what Fuller and DeCrane have said about there being people in influential areas, some within lobby groups, some within congressional staffs, etcetera, who have a real bias against the use of fossil fuels, and would like to see the country go to kinder, gentler energy alternatives, whatever they may be. But the technology is not there to compete with the fossil fuel industry.

No matter how you slice it, the best and cheapest source of fuel—I think way into the next century—is going to be fossil fuels. Other people seem to be stacking the deck against the industry. I agree with them on that. But they're talking to the wrong people. They keep talking to the choir, and they need to get out and start winning some hearts and minds on the firing lines at the grassroots levels.

Listen to the working man

Instead of sending out company executives like DeCrane or Fuller, they need to reach down to get a working man or woman from the oil patch or from the refinery who can talk to other working people and say: "This is what happened to me because of this. This is wrong. You're hurting our families. You're hurting our jobs. My kids aren't going to the college I wanted them to go to. We've had to sell the house. I've been laid off. You're hurting America because you're eliminating these jobs and my job is going overseas and somebody else is going to be doing my job in Africa or Asia or Russia or whatever. This is wrong." This industry can make the point very well that it's lost more jobs than any other major industry—more than the steel and auto industries.

It's an us-against-them mentality. You get a more favorable hearing in the Sun Belt and western states than you do in the northern states, because that's pretty much how the split is on the energy deposits. But people sympathize with the small farmer who is being driven under by the big farm operations and by federal regulations. Well, the farmers are subsidized by the federal government, and, really, the small farmers, from what I've seen on the statistics provided, are not as efficient and the cost of food that they bring to the market is higher than the big corporate farms that people rail against.

You look at the oil industry and exactly the opposite is true. The small independent can go in and take a lease that a major company is losing money on and can make a damn good profit on it, because he's there with it and knows how to work it. He's lean and mean and gets out there and really scrambles on it. He can bring oil and gas to market at a cheaper price oftentimes more than the majors can.

Hurting the "moms and pops"

Yet, when they legislate something, they aim it at Exxon and they kill off small independents—a lot of mom and pop operations, a lot of little old ladies who are collecting royalty checks. They're the ones who get shot in the head every time somebody takes an aim at Exxon. Exxon is going to survive. It's got overseas operations and other things to fall back on. The small independent who does most of the drilling in this country and who's discovered most of the oil and gas deposits is an endangered species.

I was at a meeting of Indonesian officials here trying to encourage more U.S. investment in their oil and gas system. In 1994 they eased some regulations to encourage investment.

I kept hearing that the Indonesian government will listen to the problems the industry has. If they outline a production-sharing program that the industry says is too stringent, that doesn't give them a chance to recover their cost and make a profit on their investment, they're willing to listen. They may not give them everything they want, but they will try to work something out. The same thing with Nigeria. Nigeria has worked out deals to improve returns for deepwater exploration and development.

Britain has long had a reputation of working with the industry to adjust the tax bite. If the prices are high, the tax bite is higher. Everybody, the government and the industry, reaps benefits from the higher prices. But if prices drop, the British government will lower the tax bite to give companies a chance to keep operating and to keep maintaining a profit. This has served the British very well. A lot of these countries are trying very hard to do this.

The one exception is the United States. You don't see anything really meaningful coming out of the United States. The Clinton administration's gas/oil initiative did nothing. It's all smoke and mirrors. There's very little in there that really benefits the industry as a whole and damn little or nothing, really, that benefits exploration in this country. Therefore, you've got more and more incentives for companies to go overseas and you're seeing not only the majors, but independents like Apache, doing things elsewhere. Hondo Oil & Gas [run by retired ARCO Chairman Robert O. Anderson] risked its whole fortune on a well in Colombia.

They're going where they can get the biggest bang for their buck. Where they've got people who they feel want them and will work with them. They're not going to pull totally out of the U.S. They're going to do what they can do well here, but you're not going to see the type of exploration drilling that you need to maintain domestic supplies. You're going to see a continuing fall-off in domestic reserves in both oil and gas, and more operations overseas, more imports. On the negative side, that's the biggest impact.

Technology explosion

On the positive side, what's working best and what probably will determine the survivors not only for the rest of this century but well into the next century is technology. The survivors are going to be those companies that have got the right people who can make the best use of technology and aim it at a particular niche that they can do at least as well, if not better, than anybody else. They stake out their market and concentrate on it.

There's a lot of technology developing out there. The deepwater technology, subsea production, 3-D seismic, it's been a bombshell of technological development in just the last few years. They're doing things at water depths that were the equivalent of a rocket trip to the moon just five or ten years ago.

It's true you've got fewer rigs working nowadays, but the ones that are working are drilling more holes; they're having a greater success rate on drilling; they're producing more reserves per well than what the average has been in the past.

It's because of technology, primarily, but also because of talented people. The ones that are in the industry now, for the most part, particularly the ones who are working on the rig crews, are the cream of the crop. They're the people who really have survived for years and know what they're doing. You see that in a lot of companies, like Earl Swift of Swift Energy, some of the other independents—Ray Plank [chairman of Apache], George Mitchell, [chairman of Mitchell Energy & Development Corporation].

These people have a memory of what happened before, and they're not going to overload themselves by risking everything. They're going to spread their risk among a lot of different prospects, partnering with other producers on some things. The whole idea is concentrating on what you do best, getting back to the basics, work the areas that you have the most expertise in. These companies are going to survive and prosper.

It's a totally different way of doing business now. A lot of it makes good sense. There are a lot of things that should have been done a lot sooner—the good old, bad old days when everybody was throwing money at the problems. "I need a rig. I need it by next week. The quicker I get that rig, no matter what the cost, the quicker I could be making money." That was all people looked at.

Partnering

Once that gravy dried up, they had to start looking at what could they do to cut costs. You can only fire so many staff people, so you start looking at different ways, some of which are through various combinations. They're getting away, and this is good, from the adversarial approach between the producers and the oil-field service companies.

For a long time they met across the table as enemies. "I'm trying to sell my services for the highest price I can get. He's trying to buy them for the lowest price he can pay me." That type of conflict was the natural way of doing business.

Now, they go in there and they say, "Look, if you can guarantee me so much work, then I can lower my cost to you. I can keep my crews busy. I can keep my workover rigs working. I can keep things up." So the producer says, "Here's a company we've worked with before. We know they can do good work. They're dependable. Let's throw them a little bit more business. Let's pay them a little bit more so they can be around."

With the majors paricularly, you're seeing more outsourcing for services, and that leads into partnering operations. It used to be, "You want my rig, here it is. Tell me where you want the well done, the specs you want me to drill, and I'll drill it for you. You pay me on a daily basis." Now the drilling contractor often has to come up with the way to design the well, too.

They're doing more and more of what the operators used to do in-house. And the companies are letting them do this—insisting that they do this—because it lowers their costs. Maybe the service companies are not quite yet at the point where they share as much reward as they are being asked to share risk in every case. Still, it is an improvement and it makes sense. You look at it now and you wonder why they couldn't have been doing this all along. It was a very cutthroat business for a long time.

Continued shakeout

Companies that were household names have disappeared. The Baker Company and the Hughes Tool Company are Baker Hughes. Cameron Iron Works was a great oil-field equipment company and has disappeared because it went into Cooper, and now Cooper has sold off its oil and tools operations.

A lot of very good companies are gone, not because they were bad companies, but they've disappeared into mergers and acquisitions, and some of this is still going on. You're not going to see the mega-mergers that you did back when Chevron took over Gulf Oil, or when Texaco and Pennzoil fought it out for Getty, or Mobil took over Superior.

You're going to see a lot of the small mergers in the oil-field service industry, particularly the drilling industry, which still needs a big shakeout. There are too many drilling contractors out there with too many rigs competing for too few jobs. That's bound to happen. But companies like Noble Drilling Corp. are doing their best to take a lot of those rigs off the market. You can see more shakeout in the workover industry, too.

You'll see some companies getting smaller or disappearing. You'll see some companies are going to be getting bigger. A good example was Texaco selling more than half of its domestic oil and gas fields and getting smaller, but Apache buying them and getting bigger. Apache has accumulated a lot of good properties.

It's a totally changing industry. In the long run, it probably is good for the industry. In the short run, a lot of people are going to get hurt. You've had a lot of people laid off in this industry, and you're going to have a lot more to go. In other cases, it isn't that bad. Texaco, with the sale of its properties, is cutting several thousand jobs worldwide. Some of those people who worked those particular properties sold to Apache will go to work there. You see a lot of that. You'll see a lot of people who took early retirement coming back as consultants, sometimes even to the same companies they used to work for.

Amoco has undergone this major restructuring from three main companies to seventeen operating units, where each of them will be responsible for their own strategies and profits. You plan it, you make it, and you're accountable. You've got more responsibility, but you've got more accountability, too. You're going to see more and more of that.

Swift Energy follows a very good policy of trying to judge when is it cheaper to buy the oil and when is it cheaper to drill for it. There are several companies that have done very well in the mix of drilling operations and exploration acquisitions.

When the downturn came and people were laying off workers, Anadarko Petroleum Corp. kept its people. [Chairman] Bob Allison said the investment in his people was the best investment he had. That's paying off now as they are becoming more active. He's got very good people. The morale is good in that company. It's doing well.

The Rowan Companies had some top-notch rigs and experienced crews. Rowan was losing money, but [Chairman] Bob Palmer kept the rigs working and in good repair, and kept the experienced crews together. That's paying off for him now.

Companies are much more sophisticated now. They are hedging their losses on the futures market. They know about exchange rates. They know how to write contracts, and if they don't know, they can find somebody who can tell them very quickly how to do it.

Overall, it's going to be a leaner, more efficient, more competitive industry. The ones that survive are going to be some of the best companies in the world. And they're going to be very strong companies.

The buying craze

The boom was a disaster in retrospect for the oil and gas industry. It brought in a lot of people and investors who did not know what was going on. All these people got burned.

I remember very well at an annual meeting back in the early '80s, where the shareholders of a company were in a tizzy because the administration had accumulated all this cash in reserves, and it was not spending it to buy something else. There were people standing up and demanding, "Why don't you buy something?" Like Mobil bought Montgomery Ward, and another major bought an electric power company—both turned out disastrous for them. The executives insisted, "But this is what we do best. We don't like to expand into other areas."

They were browbeaten by their shareholders into an expansion program that, when the boom turned to bust, they were left with this debt, with this extra production capacity that they no longer needed. It was a disaster. The company administration's gut intention was to do the right thing, but the shareholders pushed them into doing the wrong thing.

You had this big buildup during the boom. Around '79, '80, before everything really went haywire in '82, drilling contractors were complaining, "Damn, there's too many rigs being built out there," and "Rig rates are going to go to hell in a hand basket."

I'd say, "But don't you have a couple of rigs that are being built now that are on order?" "Oh, yes, but I need those rigs. It's these other guys building rigs that are building too many." Some drilling contractor's got ten rigs and six of them are working, and he puts two more to work. It's a boom. Same contractor, he's got ten rigs working, two come off contracts, he can't find a job for them. It's a bust. It's very subjective.

What was different about the industry back then was they had more optimism. But they've been so negative for so long now that it would take a sustained period of price stability for oil and gas to break that trend. Not necessarily higher prices, but stable prices that you can bank on. The futures market has helped destroy that, because everybody now adjusts prices on at least a daily—and sometimes it seems like an hourly—basis, on what the futures prices are. When you've got the type of fluctuations you've seen in prices the last few years—a lot of knee-jerk reactions to things—it's very hard for companies to plan ahead.

Much of the shakeout has been completed, but you're also into a very changing market now. It's a very international operation and more competitive on a world basis than ever. Part of that is because some of the companies have been driven overseas to do more work. The independents previously had no reason to go overseas, because there was plenty of work and money for them in the United States. They could do everything they wanted right here. The bust came, and they're looking around at, "What can I do, because I'm damn sure not surviving here." They've moved into areas like Latin America and Russia.

Competition everywhere

Some companies have seen most of their sales come from overseas in recent years. That's been a big change. At the same time, you've got overseas companies, many of which were once state-owned and stayed at home, competing more here for U.S. oil and gas reserves. BHP from Australia has offices here. Nippon Oil from Japan has offices and production in the United States. Petrobras, Brazil's state oil company, is working in the Gulf of Mexico. The Saudis have part of the refining operations with Texaco. The Venezuelans have invested in U.S. refining and marketing operations. British Petroleum is very big on the North Slope and has some of the biggest holdings in the U.S. of any company.

You've got service companies and drilling contractors, others that are working here in this country, that used to be overseas. On a regular basis, I see the Indonesians, Colombians, Mexicans, Brazilians, and others coming into Houston competing for the expertise, capital, and technology that U.S. companies have. "Come to my place—we will give you the best deal on your operations down there."

Argentina's a good example—wide open, a country that before you couldn't buy your way into. Today, they want people down there. Venezuela is opening up more and Mexico will, too, though not giving up ownership of their oil and gas deposits. But you're going to see more countries like that opening up, because they recognize that this country and the U.S. oil and service companies have the knowledge, technology, expertise, and money, to the extent that they need.

I've heard people argue that it's easier for a foreign producer to operate in the United States than it is for a U.S. producer, because they get a better tax rate. Oftentimes, particularly in the case of some Japanese companies, because there's a heavy subsidy by the home government, they can come off a lot more profitably because they are only taxed like any other business in the United States—whereas back home they're getting certain subsidies for their operations overseas, because they're providing supplies that their country wants or they are getting certain technology into the market. Or if they work out a deal, for instance, in some cases, to develop a field, build a pipeline in Latin America, a certain amount of that pipe is going to be made by the steel industry in that company's country. It behooves their government to give them the subsidies. It helps create jobs at home. In that sense, a lot of these companies can do better here than U.S. companies can.

Reason for write-offs

In the end, looking at all the tax "loopholes" that were closed in an effort to generate more federal funds, that probably hurt the industry more than anything else, more even than the windfall profits tax. When they closed all the loopholes, when they did away with the depletion allowances and the write-offs for tangible drilling costs, that took some real money out of this industry. Worse than that, it took the investors out of this industry—people who before would invest in wells, put money on the line because win, lose, or draw, even if it was a dry hole, they'd get a tax write-off and benefit overall.

I maintain that if you looked at why these special tax write-offs were put in there to start with, there were good reasons for it. The idea was to encourage investment in an industry that was generally viewed as a vital one to the security and economic welfare of the United States—that is, oil and gas. Oil and gas fuels everything else. You can't have a steel industry without oil and gas. You can't have an automobile industry without oil and gas. You can't move your farm products to market without transportation fuels.

It really hurt an industry that was vital, and it was done piecemeal without looking at the cause and effect. One of the worst things the government has done is the steady chipping away at the things that encouraged investments and exploration, that allowed people to go in and gain access to areas that were then frontier areas where they could work.

It took money not only out of the oil industry, but out of landholders. This is the only country where the people—individuals—own oil and gas reserves, the minerals. Every other country you go to, it's owned by the state. So all the money from exploration and production falls into state coffers. It does not do so here. It goes indirectly through taxation into government coffers. But, for the most part, other than what goes into the school fund from royalties paid to the state on state lands or the federal royalties offshore on federal lands, it goes to people.

EDWIN S. ROTHSCHILD

energy policy director of Citizen Action

Edwin S. Rothschild is the energy policy director of Citizen Action, a Washington-based federation of thirty-two statewide citizen organizations with nearly 3 million members. During the past twenty years, he has held numerous positions in the energy field, including analytical, public policy, and public affairs assignments.

Rothschild is one of the leading experts on oil and energy issues, both domestic and international. He has written and testified extensively from a consumer perspective. Although his outspoken views are not always appreciated by the industry, he is respected for his knowledge and moderation.

Rothschild believes that while the industry feels it is overregulated, this state of affairs is nevertheless what the majority of the public wants.

I don't think there's one petroleum industry. There are several segments of it, and they each have a series of different problems facing them.

Let's separate them out in the following way.

There are international companies that are vertically and horizontally integrated. Companies such as Exxon, Chevron, Mobil, Texaco, whose view of the world is that they see themselves as transnational or multinational companies. The fact that they're domiciled in the United States is totally irrelevant.

Their view of the world is much different from that of the domestic oil-producing companies. And that's a second category. There are companies that simply are in the production of oil and gas in the United States.

And then there are companies which are primarily U.S. companies, but are integrated companies that produce, refine, and market petroleum and do so primarily or exclusively domestically.

With that said, the international companies probably are in the best situation because the world, first of all, is not running short of oil. It simply is an issue of where it can be found and at what cost. The United States, as a producing province, is declining substantially with respect to oil. Not with respect to natural gas, however.

The threat to the industry is the fact that the United States is running out of oil. We've been exploring this country far more than any other country. We were blessed with a certain amount of reserves. And the likelihood of finding any large new fields is very, very limited. There seem to be some fields in the Gulf of Mexico in deeper waters. There seems to be some effort to recapture some older fields and improve the producibility of those fields. But in terms of finding big new fields, it's not very likely.

As that occurs, the larger companies—international companies— are going overseas where the costs are lower, the risks are lower, and the returns are greater. So they go to places like the Middle East, Vietnam, Indonesia, Russia, and other former states of the Soviet Union.

Domestic integrated companies are going to be vulnerable with respect to their declining resource base, and a number of them have decided to make changes. For example, Sun several years ago spun off its production side into an independent company called ORYX, and it is now trying simply to be a refiner/marketer along the East Coast, particularly in the Northeast.

So companies that are going to refine oil are likely to be in better shape than companies that are going to be dependent on oil production. I think that the prospect for domestic producers—domestic oil and gas producers—is far more tenuous, particularly if they do not have low-cost reserves.

There's not going to be any hope, I think, of finding federal support for the domestic oil industry in the form of an import fee or tax breaks. And producers cannot make money on domestic resources if they don't have oil resources that can be developed for about $20 a barrel. At $20 a barrel, those high-cost resources aren't going to be developed anytime soon, and those producers are not going to be able to continue their operations.

On the other hand, the degree to which producers are emphasizing natural gas will provide better opportunities, because I feel the gas market will expand.

One problem that arises is that global warming seems to have resumed. And anyone who has spent a winter up in the Northeast and other parts of the country will know that a lot of gas and heating oil that would have been consumed hasn't, because the weather has been so mild and balmy. That could pose a serious concern over the long term for producers. As they develop supplies that may not be consumed, the market may not grow as much as some have anticipated.

To the degree that gas is a substitute for oil, that can mean more competition. Gas, first of all, burns more cleanly than oil. If you compare the burning of gasoline versus natural gas, the end result is far fewer pollutants in every major category for gas that is consumed versus gasoline.

Attack on natural gas

The oil industry is trying—and this is clear if you look at the American Petroleum Institute or some of the companies in their advertising now—to denigrate, discredit, attack, undercut any effort to promote alternative fuels, particularly natural gas and electricity, particularly in the market for vehicles, because the major oil companies are clearly hoping and planning that the liquid fuels industry will continue to dominate the transportation market.

It looks as if that will occur in the current Congress. I don't think the new Republican Congress is about to extend what occurred two or three years ago in an effort to expand the role of alternate fuels.

I think what we'll see is a rolling back of some of the environmental regulations and laws and a renewed willingness to support efforts to maintain the use of liquid fuels, primarily oil-based fuels in the gasoline market, which is the most profitable segment of the oil market that the major refiners have. I think that in the transportation market, oil is going to continue to remain dominant for the foreseeable future. I don't know if there's such a thing as an unforeseeable future.

If the oil refiners are successful, natural gas will continue to remain in a niche market with very little growth, and there'll be a lot of effort to reduce it even further. The only growth potential that exists is in the electric utility market and industrial market. And that's clearly what they [major oil companies] want to pursue, because those are large sales to large users over a long term. That is precisely the kind of market that the large companies want to have.

Now with respect to smaller companies, the natural gas market will be basically what it is today, I think, with some limited growth. If you don't push competition with gasoline, it's going to be difficult to see substantial growth in the future if there is again a backsliding on the effort to increase efficiency in the electric utility area. And I think we may see that with the new Congress rolling back some of the advances and some of the funding for programs that support increases in energy efficiency. Because oil prices have basically stabilized since 1986, the growth in investment in energy-efficient plants and equipment has slowed as well.

So my prediction, for what it's worth, is that oil will continue to dominate the energy market. It will dominate the petrochemical market. It will dominate the market for transportation, too. That's likely to happen, in my view.

Natural gas will become more important in the electric utility area, and that will be the major growth area. They'll try to also substitute gas in the residential heating market. Electricity is certainly not as good as natural gas, nor is heating oil, in many respects. I think the competition there for new housing construction will still also be dominated by natural gas, but that's not a huge market.

No overregulation

I hardly think the industry is overregulated. I don't think it's ever been overregulated. The problem in the industry has nothing to do with regulation. It has to do with the ability of some people to compete in the market. When the price goes down because the weather is warm, that doesn't have to do with regulation.

The irony is that because of a supply surplus, a lot of producers in the states of Texas, Oklahoma, and Louisiana have run to their state commissions to get the commissions to provide the legitimacy to reduce production in order to stabilize price.

So the question is not whether there's too much regulation. The question is, What kind of regulation is there? We have different federal and state laws that impact on the production, transportation, and distribution of oil and gas. There are no price controls on oil. There are no allocation controls on oil. The only kind of federal regulation remaining has to do with safety and the environment, primarily.

On the natural gas side, there is no regulation on the price. The pipelines are no longer really regulated. And, again, you're talking about regulation having to do more with siting a pipeline than anything else. So I don't know where people can conclude that there's too much regulation.

Keep it clean

I think the public wants clean air, clean water, and clean or unpolluted food. And I think the public realizes that there are trade-offs. And just because there are companies that have been used to shifting the cost of pollution onto the public, now that we have rules requiring the internalization of those costs, the companies are unhappy with it. It's unfortunate, but that's the way it ought to be.

We ought not to make it free of charge for someone to pollute the air or the water. There's a cost involved in that which ought to be internal-ized. If that means prices for products have to be higher, so be it. But the industry shouldn't be complaining when people want to have clean air to breath or clean water to drink.

The industry only cleans up its messes under pressure from state and federal governments as the result of laws that have been passed under pressure from the public to do so. There's no voluntary industry effort to clean up its mess. The industry has to have its feet held to the fire in order to do the job.

We've seen, for example, the head of the American Petroleum Insti-tute speaking at their annual meeting in 1994, saying what a wonderful job the industry did in manufacturing cleaner gasoline. Well, the only reason that was done was not because the industry wanted to do it and advocated and lobbied on behalf of the Clean Air Act. In fact, they did just the opposite. The only reason it was done, the only reason the com-panies are making cleaner gasoline, is because they were forced to do it.

It's very hard to measure a reasonable balance [between a clean envi-ronment and industry's needs], because we're talking about components that the industry feels are too costly, and it means that they will have to comply, and it means money that they spend on something other than production to earn a return on their investment.

Promote gas

On the other hand, the public sees the need to reduce production and pollution. And the point is, that if you're in the oil business, you've got a very, very polluting product. And if you're in the gas business, you've got a much less polluting product.

It seems to me the incentives ought to be on the side of promoting cleaner burning fuels. I think if the industry wants to do something and wants to work with consumer groups and environmental groups to pro-mote cleaner-burning fuels, they'll be promoting natural gas and not oil—particularly independent producers who are primarily producers of natural gas as opposed to oil, following the lead of people like T. Boone Pickens, who think that it makes sense to promote natural gas and not oil.

Natural gas is what we have left in this country, and we shouldn't be importing all that oil from overseas under the auspices of some of the larger companies that are international in scope and have little reason for worrying about the domestic viability of the independent producers. These major companies are concerned about promoting oil as a product.

Efficiency standards

I would advocate a strategy that emphasizes investment in energy efficiency. That is, in all of the equipment that we use, we ought to have programs that emphasize the purchase and/or requirement that certain products, whether air conditioners or heating systems or whatever, meet a certain efficiency standard. And that's anything that uses energy, whether it's automobiles, lighting, refrigerators. That's number one.

By the way, that would promote domestic investment in products that we could then sell both here and abroad. That's what the Japanese and Germans are and will be investing in—new technologies. There's a large return for that, both in economic growth benefits and job benefits.

Secondly, I would emphasize cleaner-burning fuels, renewable fuels—solar, wind, and natural gas. That ought to be the direction we're going. We've seen the high cost of nuclear power.

We've seen the high cost and polluting consequences of the use of coal. And we've seen the high cost and polluting consequences of burning oil. And the high cost of sending or keeping an armed force ready to protect the sea lanes from the Persian Gulf. And the high cost of paying for imported oil. Not the national security cost so much as the economic cost.

For all those reasons, we ought to be emphasizing an energy policy of efficiency and cleaner-burning fuels.

Cost of drilling

You have to balance the industry's demand for increased drilling against the environmental and ecological consequences. The industry has not shown itself to be particularly trustworthy in this regard. And if you took a look at Alaska, the industry has an abysmal record with regard to the Trans-Alaska pipeline. There are serious environmental problems.

And when the price of oil declined, a lot of those companies cut back on their environmental investments. So, I think those concerns are limited.

There's one area, primarily one area, that the major companies want to open up, and that's up in Alaska. And the point is, there are other things that can be done. We don't need to open up every area to exploration. There are other values to be emphasized. Drilling prospects in the offshore Gulf of Mexico are being developed. That's not closed off. There are some areas off the West Coast that are still being drilled. So, the point is, if the industry can't get access to everything, that's not a terrible consequence.

In addition, you still will have whatever oil and gas is in the ground. It's not going to disappear. It's not going to go anywhere. It's not going to migrate to some other country. So, if in the future, there may be more concern or more reason to do it, it's still there. And we don't have to do it at present.

The people's government

Government's role is whatever the majority of people want the government's role to be. There's some misconception that the government is something other than what we have basically created.

It seems to me the government's role ought to promote the public's best interest. And at a time when we're facing a variety of concerns, whether they be economic or environmental, we must try to find some balance. And the balance here is to promote, in our view, some new industries that will shift us away from polluting fuels to less-polluting fuels.

That doesn't have to be a costly transition. We've made transitions from other energy sources in the past. We went from wood to coal to oil. That doesn't mean that oil is inevitable. People have to realize that there may be other avenues to consume energy than simply consuming the fuels that we have in the past.

And that's why we think the government's role here is to develop and support research/development efforts that the industry will not, particularly in terms of efficiency and renewable fuels. To provide incentives to some of these newer industries to put them on the same footing as existing industries will benefit the nation in the long term. There's no way a new industry developing solar cells, for example, can compete with a mature industry which has sunk costs. There's no way. So that's another role for the government.

And, finally, in terms of protecting the environment, companies by their own logic, in a free marketplace, do not and will not curb their own pollution. That has to be a role of government. To set a standard for industry and to enforce it.

Innovation will lead

Leadership will be assumed by companies that are innovative—those that aren't tied to ways of doing things in the past, ones that take advantage of the public's concern about the environment and recognize that that has to be included in anything they do. If they figure they don't have to worry about that, they're wrong.

They also have to recognize that there are a lot of opportunities, but that there is a real conflict between companies that really want to compete and are small or medium-sized versus those that are large. Larger companies tend to dominate the marketplace and make it very difficult sometimes for smaller companies to survive during periods of change.

So, if you're going to be somebody going forward in the oil and gas business, you have to realize that in order to survive, you cannot expect, one, that the government is going to help you out very much, and, two, that you shouldn't expect the price of oil in real terms to go much above $20 a barrel.

If you want to survive in the oil and gas business, I think it makes sense to focus on the exploration and production of natural gas. If someone in the oil and gas business wants to stay in business, one, I would say gas is the way to go, because it's both environmentally more benign than oil, and, two, you're probably much more likely to find gas in this country, because the resource base is still quite large, and, three, I think the public is much more willing to see gas used as a fuel as opposed to oil, because in the longer term, oil is still going to be much more vulnerable to criticism.

I suspect there is a different type of energy executive today. It depends, again, on what level. Some people have to be more attuned to developments not only in their own industry but in others. They have to be attuned to rapidly changing technology. They have to be attuned to changes in the workplace. I think the day of the individual entrepreneur in the oil industry—going out and finding oil and gas—is relegated to our past. The future looks like you have to be part of a larger organization in order to stay in business.

Imports

I don't think the volume of oil that we import is the issue per se. The Japanese import 99 percent, they've succeeded economically. The issue really is what effect it has on our balance of trade. What effect it has on the use of other fuels. The problem in this country, primarily, is that people think that oil should be cheap. We've grown up with low-cost gasoline compared to people in Europe and Japan, where a gallon of gasoline costs $4. Here it's kept quite low. In fact, the tax on gasoline hasn't kept up with inflation. The price of gasoline now is cheaper than it was thirty years ago.

So, first of all, we're dealing with an inherent belief that we have to have low-cost oil in order to survive, which promotes the importation of oil. And unless something is done about that, imports will continue to rise.

Most people in this country don't know that we import that much oil. Most people don't care. The only people who care about this issue are independent domestic producers, and they are constantly making an effort to get the federal government to define this as a national security issue.

They will be unsuccessful because—and here's the rub—too many people in this country—both consumers and big businesses that require oil—are not going to support anything that artificially props up the price.

Government

JAMES WATKINS

secretary of energy, 1989–93

Just two months after retired Admiral James Watkins took office as George Bush's secretary of energy in 1989, he was confronted with the ultimate challenge: the *Exxon Valdez* disaster in Alaska. The offshore incident cost Watkins, a former chief of naval operations, many sleepless nights and destroyed any momentum the petroleum industry had established in trying to improve its image. Despite the setback, Watkins pushed ahead against all odds to see the enactment of the Energy Policy Act of 1992 in the last days of the Bush administration.

Watkins fought opposition from environmentalists, influential members of the petroleum industry itself, and fellow members of Bush's administration in his effort to see this nation enact some sort of energy policy that provided relief to the independents. That he did so speaks of his determination, his ability to organize, and the close relationship he had with Bush, who was one of Watkins' biggest admirers, according to friends of the former president.

A graduate of the U.S. Naval Academy, Watkins has a master's degree in mechanical engineering. He is the current president of the Washington-based Joint Oceanographic Institutions. Watkins was more than willing to share his thoughts about the petroleum industry, which he discussed with intense passion and frankness.

*I*n my mind, the key issues are fundamentally unchanged from the ones I faced at the end of my four-year tour in 1992 with the Department of Energy. I certainly felt they existed then. I feel they still exist today and are going to plague the industry and hurt our economy unless we address them forthrightly.

I put the number one issue as uncertainty concerning future directions of both national and state lawmakers regarding access to certain of our nation's natural resource treasures—like the Arctic National Wildlife Refuge, so-called ANWR, and the Outer Continental Shelf— our resources for both oil and gas. This is very critical.

The analysis we conducted during the Energy Policy Act of 1992 debates indicated that development of new U.S. oil and gas resources would be essential over the next ten to twenty years if we were going to limit the kind of economic damage this country has already suffered— and some of our allies around the world have suffered—from another threat of Mideast oil cutoff. After all, that threat was one of the primary driving functions we were to address in the National Energy Strategy development, which we did.

We felt that while we could wean ourselves off of oil in the transportation sector to a certain level, we had to come down gracefully, recognizing that technology had not matured to the point that we were ready to make a rather sudden, guillotinelike switch to alternate sources of transportation fuel. It would take time. We predicted that we would only have about a million electric cars and compressed natural gas vehicles on the road by the year 2000, out of nearly 200 million cars on the road at that time.

Realistically, we have to look at how to transition from the insatiable oil appetite that Americans have in the transportation sector to some alternate fuel. Theoretically, I suppose, you could step into these other alternate sources overnight. But technically, realistically, and economically, you cannot. So we still need traditional fuel resources for a long time to come.

I'm hopeful that the recent conservative political revolution that just took place in our country with the November 1994 elections will allow the debate to be reignited. For example, we need to reopen the debates on ANWR and the Outer Continental Shelf. Those are both critically important, not for forty years and beyond, but for the next twenty. Let us wean ourselves away from the potential economic impact of another Mideast crisis as best we can. Let's do it in an environmentally sound way with the latest technology. Let's keep our eye on it—keep a congressional watchdog group if you want to—on the environment. Let's do it right and get those economic benefits back here, instead of sending them overseas.

Let's get ready to pull the gas out of the existing Prudhoe Bay fields. Let's get those pipelines ready to go. Let's get liquefied natural gas [LNG] produced and exported to Japan out of Alaska. Let's get our contracts going now, so that we can begin to build a better transitional mechanism for movement away from our current dependence on oil as we move into the next century. I think we need another twenty years to do that. That's about as long as it's going to last anyway.

High cost of sitting still

For this to happen, however, we need to work with the states and demonstrate to them that these natural resources can be utilized in an environmentally sound way. If we don't do this, if the access continues to be disallowed, the industry will continue to be forced to protect their shareholders' interests by taking more and more of their business abroad. What happens then? Crude oil imports to the U.S. will continue to rise in percentage of total oil consumption. Such imports will aggravate the already burdensome trade deficit situation we're trying to eliminate.

Moreover, it's not environmentally sound to import more oil in tankers. It's far better to pipe it in from offshore. Our offshore drilling excellence over many years has demonstrated that we simply don't put very much oil into the ocean in the process. In comparison with the natural seepage of oil into the oceans, what man adds is trivial.

The strategy we are now employing in this one area is basically a lose-lose strategy. We lose all the way around: environmentally and economically. We're losing because politicians find little personal benefit in confronting extreme environmentalists, who are really driving current policy. That's sad, because there is no substantive justification for this extremism. Yet politicians acquiesce to it. That's what we saw on Capitol Hill: our inability to put common sense into the equation during the debate. That's my first item.

Competitiveness threatened

The second item that I feel is extremely important is the uncertainty over how far the future imposition of environmental regulatory controls will go. If today's trends are not dampened to some extent, it could render the industry even more noncompetitive with foreign suppliers of oil and liquefied natural gas. If we can't come to grips with a realistic risk assessment of what we are really doing to the environment as opposed to what the allegations are by these same extremists, we're not going to be competitive.

We could find ourselves importing refined products as well as raw products. For example, if we can't build new refineries in this country and if we're limited to the extent we are today by what I consider to be excessive environmental litigation and misuse of the intent of the various laws on environment, then we will certainly do increasing damage to the trade deficit and to our ability to compete.

The worst thing we could do, in my opinion, is import liquefied natural gas in large amounts. Why? We have plenty of gas reserves in this country. But environmental extremists believe that somehow you just turn a spigot to get gas. You don't drill for it. We know how to drill for gas. We know how to drill in an environmentally safe way, and we should be allowed to do it and bring some of our industry back home that's now overseas. Uncertainty over future regulatory excess is very real and raises serious questions again as to whether we can overcome extremism with common sense.

The third item is another uncertainty: the future value of oil on the world market. One big issue in the energy strategy development was that nobody could really project what oil prices would do over the next decade. Yet the world price of crude drives so much of the economic projections. How much will OPEC be able to charge for the next ten to fifteen years in the world market? With the passage of GATT [General Agreement on Trade and Tariffs], we expect nations to try to open up their markets. Can OPEC or Mideast nations, for example, have it both ways: free market when it suits them, controlled market when it doesn't?

That means a lot to the United States in terms of our making meaningful economic projections and what we can count on regarding cost effectiveness for oil production here. Absent this information, it is virtually impossible to assess the economic competitiveness for alternate sources of energy for the future. This is another critical concern and contributes again to uncertainty.

Technological opportunities

The fourth issue is what I would consider the future potential of technology and deployment to meet our growth needs. This is further aggravated by the increasingly stringent environmental demands that could be imposed and become insurmountable barriers to meeting these needs.

Here's an example: Let's say we can develop new technologies for either enhanced oil recovery from existing fields or expanded ocean recovery from new finds using new three-dimensional characterization techniques that computering now permits. The National Laboratories can help by sharing their more sensitive and high-tech seismological equipment with our oil industry, particularly the mom and pop independents that give us so much of our oil but don't have the resources themselves to fund this kind of research. There is little question that we can enhance their recovery by helping them work together with others.

Perhaps both enhanced and new oil recovery can add to the ANWRs and the OCSs and give us time to transition more gracefully to alternate forms of energy.

On the other hand, if those technologies are not allowed to be deployed or are constrained because of environmental requirements, or if there's no incentive because oil prices are so low that companies can't keep their fields working, we will again see our industry move abroad in order to protect their shareholders.

One clear conclusion from our debate on the Energy Policy Act of 1992 was that the nation cannot disregard the important role that oil and gas will play in the national economy over the next twenty years. Natural gas now is moving into our transportation sector. It's moving even more substantially into our utilities—power generation in new ways—all of which are going to require more gas production. We've got to get serious about the sources of gas and its availability. We have a large supply, so let's get serious about it. The Energy Policy Act tried to find the balance by showing that with a good integrated energy strategy, oil, gas, coal, and renewable sources of energy would all be necessary to ensure a robust economy for the long haul.

All of these uncertainties were there when I was secretary of energy, and they continue to be there. My only hope is that there may be a saner view being taken by the new leadership in Washington that will open up this debate again, and get much more sensible about the mix of energies that are necessary for this country to maintain a 2.5 percent to 3 percent GNP growth per year and continue our competitive edge in the world.

Without the government being actively involved, with the regulatory controls that are imposed, and with the inability to drive world economies anymore (we're a small percentage of the world's production of oil), we have to be very cautious. I'm not saying we shouldn't be free traders, just that as we transition from the situation that has pertained since the early years of oil and gas production to one that wants to wean itself away from heavy reliance on that form of energy, we must have a steady and sensible transitional mode. That can only be set up by a mutual understanding among the states, the federal government, and the energy industry.

There has to be a much more collegial approach to this. We need to get serious and transition in a sensible way, instead of getting pressured during election years by the ethanol, methanol, oil and gas, and other special-interest people. Let's do what the Energy Act of 1992 tried to do—pull all this together into a package and do it in a sensible way.

I believe the industry has been too parochial and too one-energy-source oriented, and should be applauding such things, for example, as clean-coal technology. They should be applauding alternate sources of energy for automobiles and for trucks. They should be very actively engaged with others to see what the proper mix of gasoline and oxygenates are. In many cases, they are doing that. But they should do all of this very publicly and openly, saying all energy sources are good, if properly managed.

We should be part of an integrated management structure and do everything we can to share our knowledge with others. I recognize that this is perhaps against normal competitive business practices, but somehow the image has got to be improved.

The battle of '92

Unfortunately, I don't find that there's much political interest in energy. I didn't even during the debates on the energy strategy. It just wasn't that significant to the public, except when we went into debate on environmental aspects of energy production, like ANWR or the Outer Continental Shelf. Then the debate got very political. But until we bring environmental activists into play, it's kind of a ho-hum sort of issue that doesn't seem to have political clout. It's a shame. It's like education. It's critical to our economy and future competitiveness, and yet it doesn't seem to rate very high on the political agenda. Until we have a broader view taken, I don't see that politicians want to take on any new enemies in their local campaigns.

Look how very difficult it was to get the Energy Policy Act of 1992 passed. President Bush was with me, and that's the only reason I was able to get it through. I had many people within the administration—the Office of Management & Budget and Treasury, principally, Dick Darman [OMB] and Nick Brady [Treasury]—and Cabinet members who were very much opposed to any energy strategies—who thought it wasn't necessary—and felt that I was getting into free market issues we shouldn't be getting into.

[White House Chief of Staff] John Sununu thought it was politically unnecessary, raising a lot of issues you didn't have to raise. After all, the president had a 71 percent rating. So why do all this? But President Bush wanted it, because he felt it was right. He had experience in the oil and gas industry. He felt very strongly about the independents and the support they needed. He felt it was essential. He had been in politics long enough to live through a couple of energy crises—the Persian Gulf War was a striking example of that—and he felt very strongly we needed an integrated national approach. Had it not been for him, I could not have done it. The bureaucrats were unable to persuade him otherwise, fortunately. It was the only piece of substantive legislation passed during his last year in office.

I had a lot of support from other Cabinet members and on Capitol Hill, including Democrats, because they were in charge at the time. Senator Bennett Johnson [Democrat of Louisiana] and Representative John Dingell [Democrat of Michigan] did not think we'd ever get it done at the start, but were very supportive of me and went to the mat in the final year of the four-year debate on the development of the Energy Act. This is one area again where President Bush got very little credit. When we announced in Louisiana that the bill had passed and he signed it, there wasn't any big groundwell of support nationally. And yet people complain about a nickel rise in gasoline prices.

That's where our mentality is on things like this. As I say, it just doesn't have the political clout that issues like abortion or health care have. It is a very complex area—people don't understand the regulations, transmission access or oil and gas drilling. All they want is cheap gas prices at the pump and leave them alone.

Environmentally aware

Two examples of extreme positions, I found, were the failure to allow the use of the OCS and the ANWR debate, both of which had absolutely no substance to them in terms of real environmental risk. Other examples include regulatory actions taken against the oil and gas industry to disallow certain fluids to be placed in land areas where they were clearly not hazardous, and yet were declared so by regional regulators and so forth.

We're doing a lot of damage to both the image and the substance of the oil and gas industry by utilizing political activism as a mechanism to stir the pot, so that you're not going to get much response out of the politicians other than to vote these things down. There's no penalty for doing so. So environmental extremism tends to prevail.

On the other hand, help may be on the way. I think we are going to see a rebirth of interest to open some of those issues again and try to deal with this thing in a more sane way. I would expect ANWR and OCS, along with excessive regulations, looked at in terms of their relative merit now vis-à-vis anything else that is of high risk. How high are the risks of putting drilling fluids on the desert floor? Are they really hazards, and if so, to what extent? What are the standards on those hazards? Why do we suddenly call saltwater down a well okay, but saltwater out of the well hazardous? And to what extent? Is it because it contains some natural radioactivity? If so, how much? Is it a real health risk, or are we just declaring anything that's anthropogenic or man-generated as hazardous?

Those are the kinds of things I saw as secretary of energy that disturbed me. Being an engineer with some technical background, I couldn't find any justification for it—relative to other more egregious risks—either from a health or environmental point of view. Yet extremists were taking advantage of our general scientific illiteracy in this country to scare people into accepting their ideologies. As a result, they were able to defeat initiatives which would have improved the economy without damage to the environment.

The energy companies I was associated with did an excellent job on environmental issues, I felt. I went up to ANWR and saw what the oil companies, both foreign and domestic, were doing and their attention to the environment was exquisite. They've gone to horizontal drilling and small, consolidated drilling pads. They've reduced the size of the classic footprint in an area. I watched them reproduce and rebuild the tundra in areas where pads had long since been shut down. The tundra was as much like the original as you could possibly imagine. In fact, there were some cases in which they fertilized too much and left it in a condition that was even better than nature had provided.

Their sensitivities to sensible approaches to the environment are high. They've received Audubon awards for the kind of work they do, even in the wetlands areas. I don't think they're the villains. Of course, you have an occasional foul-up in which they don't do very well—the *Exxon Valdez* case comes to mind—then we do great damage because one such visible event becomes so highlighted that it becomes reflective of the entire industry. The whole industry takes a hit. *Exxon Valdez* was grist for the liberal press mill, which in its own way implied that big industrialists were the same. Many of the top executives of the major oil companies were as appalled as we were over the poor handling by Exxon of this disaster. Somehow, mandated group therapy for all when one fails has got to stop.

But let's face it: it was a situation that was very poorly handled. We've seen it now go through the courts with successful, albeit incredible, claims. I think a forthright approach at the outset, such as Johnson & Johnson took when their Tylenol was penetrated in one of the stores, is the only way to go. Their CEO got into the act at once. He was nondefensive and said, "We're going to do better." Sure, they took a hit financially, but it was short-lived. They got right out in front. The American people had credibility returned in their minds to Johnson & Johnson products right away.

Exxon did not do that. Their leadership did not handle that well and let the entire industry down, in my opinion, by being too defensive and not facing up to the issue that was involved there. They could have easily done that. There were a lot of lessons learned out of that incident, including federal government response, which could have been better handled as well. All that's been jacked up now. Out of the entire event, better things will come. But this was a rather classic example wherein an entire industry gets hurt by another colleague in the same industry, when just one of these sensitive issues is mismanaged.[1]

---

1. Exxon Corporation rejected all requests for an interview for this book.

Paying for past arrogance

The image of the oil industry—and I'll say oil as opposed to oil and gas, because people don't always couple them together, that is, gas is good and oil is bad—is not as good as it could or should be.

There seemed to be a certain arrogance built up within the oil and gas industry in the early days. They wiped out some of the electrical metro systems that existed in the Los Angeles area, for example. They overwhelmed everybody with their trucking routes and the highway building and so forth. They rode high for a number of years. I don't believe they sensed a changing attitude twenty years ago, particularly as oil prices soared and we tried to come to grips with that.

That's when they needed to come together with their own strategy for the long term that could have been much more positive, much less arrogant, more supportive of a balanced system that was clearly coming. They have to take some of the blame. The thoughtful ones I've talked to understand that and are trying to take a more enlightened approach.

They have not stepped up to the mat on the variety of realistic alternatives for the transportation sector, the prime economic driver in this country. They should demonstrate, not only in this but in the petrochemical industrial areas, that they're extremely sensitive to maintaining a healthy environment. The public is not aware of this and, of course, the press doesn't help.

They ought to say simply, "We're going to do better. We are better. We're going to start an educational program in this country to show that over the next ten years you can count on us to be a key part of the responsible environmental movement that the nation wants. We'll be very open when we make a mistake and spend our own resources to fix our problems when we cause them. We think we can maintain a healthy environment and still exploit our natural resources in a way that is environmentally sound."

They have to find a substantive way to accomplish this objective over the near term. Obviously, almost everything we do has an environmental component which we all have to watch. So you can't put a quick fix on this thing. It has to be an educational process over time. That we don't have many petroleum engineers coming out of our universities is indicative of a declining communication link with the public.

They need to find new strategies and do this in a sensible way through the schools, pre-college, that tries to explain what the situation is, that is, the environmental technologies they employ as well as the process they will follow over the next five to ten years in parallel with what will be advanced oil recovery endeavours. To be successful, they need to lay the groundwork for future acceptance of oil as a key part of our economic growth.

That kind of strategy needs to be built. It can only be done by the industry in an honest way—not overplaying it. The public needs to see them as sitting there poised, ready to deal with any kind of a spill, tanker or otherwise. It's going to require a cooperative effort between the technology developers in industry itself. Our national laboratories are working with them through the Cooperative Research and Development Agreements. It's going to have to include a worldwide partnership where the best researchers in the oil and gas industry go after the problems on an international basis. There are precedents for doing this.

Independents to step forward

I did not see the majors and independents coming together, which I considered a mistake. I thought there could be a better alliance on these large policy issues that face the entire industry. The potential for the independents to step forward has the greatest public support in this country. I would probably have said otherwise before *Exxon Valdez*. We're probably still too close to that—the majors simply don't have that kind of credibility. But I think the independents—mom and pop drillers—do. They're grassroots people. They're certainly centered in a powerful portion of our country politically and could come together with a much greater voice and be heard. But too many of their wells are being shut off. They need help and should get it.

In ten years, I think the industry is going to be where it is today minus the degradation of our domestic oil production capability, which will be significantly reduced. Imports will continue to rise to meet our gross national product needs—the petrochemical industry included. I fear it's going to continue to deteriorate, unless we come to grips and stabilize the situation to the extent OCS and ANWR can.

We need to make more gradual today's steep decline in domestic oil production and expand gas production over the next twenty years. It's very iffy right now. The projection could be rather healthy for twenty years, or it could be unhealthy in ten.

DONALD P. HODEL

secretary of energy 1982–85, secretary of interior 1985–89

Donald P. Hodel is the only person to have held the two most important government positions affecting the petroleum industry. During the Reagan administration, he served as secretary of energy (1982–85) and the secretary of interior (1985–89). Harvard-educated, Hodel was the first energy specialist to run the Energy Department. He was widely praised for restoring morale as well as for his ability to quickly grasp difficult issues. His major accomplishments included shaping legislation that led to the deregulation of natural gas.

During his tenure, Hodel was in the middle of constant battles between environmental preservationists and pro-development interests. He also contended with influential members of the administration, particularly Treasury Secretary Donald Regan, who were insistent upon low oil prices.

Hodel is a partner of the Summit Group International, a Colorado-based energy and natural resources consulting firm. In 1994, he wrote a book with Robert Deitz titled *Crisis In the Oil Patch*, which blames the government for increasing American dependence on foreign supplies to dangerous levels. In our conversation, conducted in late 1994, Hodel warned of the consequences that would accompany a declining domestic industry.

*T*he largest single issue facing the industry is the world oil market, which is dominated by OPEC or members of OPEC. They have the ability and the will to use the price—to use price manipulation—in an aggressive manner.

The production of oil domestically in the United States is generally more costly than the production of oil offshore. That means that if the offshore foreign producers decide to reduce prices drastically, they can do so as they have done and bankrupt the small, domestic producers. Medium-sized, domestic producers suffer tremendously; many go out of business and the like.

The other substantial risk, I feel, is that under GATT, incentives for stripper wells—our small, marginal well production—will be eliminated, and OPEC would certainly welcome taking a million barrels a day of production out of circulation. Once it's gone, it's gone. The fact that the United States is leaving millions or billions of barrels of oil in the ground, really wasting it, is just unavoidable.

It seems to me the big challenge is economic from the periodic, drastic price cuts in the world market. Add to that the fact that there's a hostility toward hydrocarbon production and consumption in our society, imposing additional costs from time to time on the industry—you've got two trends running against them. If both trends run against the domestic industry, it has pricing problems and then increased costs.

The upside is the current projection appears to be that the growing demand for oil in the world is such that in another two, four, or six years, depending on the rate of growth, the production capacity of the present world producers will be fully utilized. You could even bring Iraq back on the market and still have a tight market, in which case there will be rising prices.

The rising price will generate more exploration activity in promising areas around the world—though not terribly much in the United States, because of all the environmental constraints. And sooner or later, somebody will find enough additional oil so that the price will either stabilize or start back down.

Vital industry

It's a tremendously important industry. Domestic production is exceedingly valuable and important to the United States, and the United States has this almost death wish when it comes to producing domestic oil, making it harder and leaving our industry subject to predatory pricing.

Government ought to take a look at the long-term energy future of America. It ought to, in today's language, seek a sustainable energy future. That would mean one in which we could move toward less dependence upon imported oil. Which means you have to have healthy, renewable conservation and alternatives industries. It is very hard to see how you can have that without some kind of protection against predatory pricing.

Even in an era when this country rejected the idea of government intervention in the marketplace, we said you can't engage in monopolistic price-fixing. Yet that's what we are subjecting our domestic energy to, and it's not just oil. It's solar, wind—they're all up against the same thing. They collapse when the world oil price collapses.

The government could rightly take the position that it will refuse to permit American investment in American energy resources to be destroyed by a willful, predatory pricing strategy from outside the U.S. I've got a lot of friends who say, "That's meddling in the market." I say, "Wait a minute, that's not a free market you're meddling in!"

In my book, *Crisis in the Oil Patch,* I recommend that we put a floor under the price of imported oil. I propose a sliding floor because, if world oil production reaches the point where the world price has to fall and will stay down for a long time, you can't hold a floor against a world price that is persistently lower. You have to follow the world price, but the speed with which you follow it will make a big difference on how long a predatory pricing mechanism will work. If OPEC decides to drop the price to $10 barrel and the United States says, "We'll continue to buy oil at $17.50 a barrel," OPEC is leaving a lot of money—$7.50 a barrel—on the table. How long would they be willing to hold the price at $10?

I believe the energy industry is more important to the basic health of our economy than most people give it credit for. Therefore, we ought to be sensitive to issues like attack-pricing by foreign producers, recognizing that the only reason we'll get a lower price for a while is that by doing so, they can drive our own production out of business, and then they've got us at their mercy. It doesn't make sense that we would jeopardize an underlying critical industry like energy when it is absolutely essential to everything we do in our economic activity.

40 percent share

We have no alternative to petroleum products for the foreseeable future. Take all of the energy sources. The fact is, petroleum is the big piece of the action: 40 percent of the quads of energy that we use today. Twenty-five years from now, making any reasonable projections, even assuming dramatic increases in renewables and alternatives (growing 35 percent per year, compounded), and 3 percent per year growth in conservation, if you make any assumptions about economic growth in society, we will be using a smaller percentage of petroleum out of our energy mix (about 28 percent), but it will still be a very large amount of oil (25 quads).

People for an Energy Policy put together some charts and some analysis. The way it worked was if you start where we are today and assume a 35 percent per year compounded growth rate in renewables and alternatives between now and the year 2020, and a 1 percent per year reduction in oil growth, and whatever the current economic growth rate is, you still end up with the U.S. needing to consume 10 or 12 million barrels of oil a day. That would be a 50 percent reduction, but it's still 12 million barrels a day. If we could maintain our domestic production, we'd still be lucky not to be exceeding a 50 percent import rate.

DONALD P. HODEL / 339

Any way you look at it, without a nuclear attack on our domestic economy, literally—something destroying all our infrastructure—we're going to be heavily committed to petroleum for the foreseeable future, barring an unknown, incredible breakthrough of some kind which nobody can predict.

The Russian gamble

Let's look at the Russia factor, too.

One option is if Russia fails to get its act together and its enormous production potential doesn't reach the market. That means all the other players in the world are the ones that matter.

Another possibility—Russia gets its act together and it brings in Western investment, develops its enormous potential resources, and becomes a major exporter of oil. That would have a tendency to reduce or stabilize the price of oil. It would take some of the control of the market out of the hands of the OPEC members, especially the Middle East.

A third option that has to be considered is that because of the economic failures in the former Soviet Union, the hard-liners reassert themselves—whether they are communist or not really isn't the issue—and engage once again in a nuclear arms buildup or arms threat. We are presently acting as if they no longer have those weapons. If this happened, we could face a resumption of the Cold War, or worse, which would affect much more than the oil market.

Because of the hostility of America to oil production and the high cost of finding and producing domestic oil—it's not only the government that's hostile, but it's a very mature region—people with the know-how, the money, the machinery are looking for places where there's better opportunity. Certainly, if the Russians got their act together, they would siphon off investment dollars because the people who have them say, "Where's the best place to put them?"

We saw this when Canada adopted its national energy plan. In the 1980s, the Canadians under Prime Minister Pierre Trudeau adopted a national energy plan which imposed all kinds of burdens on their domestic oil industry, and the oil industry of Canada voted with its feet.

We had an influx of drilling rigs from Canada into the United States, because they went looking for the most promising areas. And Canada was no longer promising.

The majors are going to do that anyway. They've got the ability to go overseas. The larger independents likewise. Mid-size independents will look at it, think about it, and as people learn how to do it, they may do it increasingly. But for smaller independents, it's awfully hard to deal overseas without it being a special niche of some kind. If someone has relatives in Moscow, speaks Russian, he may be able to do things that some other wouldn't be willing to. Otherwise, it's an expensive proposition. The travel is expensive. You need interpreters. You must be willing to sink a lot of up-front money into going overseas.

Also, at the present time I have not yet run into any foreign investor in any business in the former Soviet Union who has a success story. I'm not saying there aren't success stories. I haven't run into them. I've never run into anybody yet who's invested money over there and has gotten money out. It's all potential. Someday that dam will break and people will find a way and it will begin to work, and the system, hopefully, will go along with it. But at the moment, I've never run into anybody who's said, "Oh, yes, I invested a million or $10 million or $100 million and I'm getting a percentage return on investment." At the moment, it's all development money.

Idealists

I haven't seen a real abating of environmental issues, either. President Clinton has put people who came straight out of that constituency and are beholden to it in so many of the key positions. They're beholden to the ideas. They are committed environmentalists, in the extreme sense of the word.

If the Clinton administration is replaced, and you still have a Congress that is more conservatively oriented, it's possible that some of the more extreme and unreasonable environmental rules could be eliminated. That would be beneficial, even though it's not the whole story, by any means.

Regarding exploration of federal lands, the opposition is political. In my view, it is not rational to refuse to explore promising acreage for fear of what is really a temporary occupancy of the surface. I consider myself an environmentalist. Everybody I've ever run into is an environmentalist. Nobody I know has ever said, "Gee, I want my kids to grow up in a dirtier world with dirtier water and dirtier air." We all want an improved quality of life. We are all, in that sense, environmentalists.

I happen to have chosen to live in an area with beautiful scenery and trees and outdoor opportunities, and I enjoy and take advantage of those opportunities. So I consider myself attuned to those values. But I don't worship nature. I don't think nature is more important than mankind or the quality of life. I happen to believe that you can go into an area, develop this kind of a resource, restore the land when you're finished to essentially its original condition, and the world will never know the difference. Except that it will have provided wealth, which in turn will have contributed to a better quality of life for more people.

That view is totally rejected by the high priests of the environmental movement. I use the words "high priests" purposely, because really for some of the core leaders—believers in the environmental movement—it is more spiritual, more like a religion, than it is a political cause. I encountered this in the hearings before Congressman John Seiberling [Democrat of Ohio], who was an ardent environmentalist.

The issue, as I recall it, was an intrusion into a wilderness area. Somebody had mistakenly thought he was on his property and bulldozed a road into a wilderness area about a quarter of a mile. He was required by the Department of the Interior to restore the surface and replant it, and the expectation was that in five to ten years you would be unable to tell that it had been damaged, that it had ever been entered by a bulldozer.

I was pretty comfortable showing up for this hearing, and I proceeded to describe the situation. You'd have thought I had just committed a horrible indiscretion. It turns out, of course, that the wilderness is the "holy of holies." The bulldozer defiled it. It didn't matter how much repair work you ever did, the fact that it was no longer virgin wilderness, that it was defiled, could never be erased.

Now, for that kind of a person and that kind of thinking, to say, "We'll go into the Arctic North Slope. We'll put down some wells. We'll take billions of barrels of oil out of the ground and contribute to a better life for people all over the world for a twenty- to thirty-year period of time. And at the end of that time, we'll plug those wells. We'll remove the signs on the surface. We'll replant the tundra, and a few years later you will not be able to tell that man was ever there," is no answer at all. They say, "But we'll know you were there. This is no longer virgin territory."

The law right now is on the side of the extreme environmentalists, and that's a "good reason" you can't open the Arctic National Wildlife Refuge. It's hard to see how in the current political climate you could do it. The environmental community raises enormous sums of money off issues like that.

Let's suppose that a conservative president were elected in 1996 and you still have a conservative Congress. It's conceivable they could remove the restrictions and permit that to go forward. There would be a tremendous outcry from the left. Whether they would have enough forces to stop that, I don't know. They get enormous assistance from the major media. Any claim of theirs is given wide currency, regardless of its foundation in fact. They raise more money each year to spend on environmental issues than the two major parties spend on presidential elections every four years, which is really mind-boggling.

I believe that any major player today in resource development is careful and sensitive for all kinds of reasons. First, you've got many more people in every company now who have gone through what amounts to environmental sensitivity training. They've come out of schools where they've heard a lot about it. Their kids are talking about it. They see it on television. It's a thriving ethic in this country. So they just are basically better oriented toward the environment.

Secondly, there are horrible penalties if you aren't careful. The two things, together, have made an enormous difference in the way people operate. What's interesting to me is that people who are genuinely concerned about the environment and are personally involved in the process will make reasonable judgments about what makes sense.

And those reasonable judgments would sometimes permit them to do much more than they are allowed to do by law. The law is much more restrictive in many cases than makes sense because you've got a broad-brush, one-size-that-fits-all approach, instead of having it be subject to the local conditions.

Instead, it's almost impossible to obtain a consensus on energy in this country. There's very little knowledge of how the energy systems work on the part of the average citizen. People want electric lights to come on any time of the day or night, when they want them, and they want electricity to be cheap. But a significant number are perfectly inclined to say, "Don't build any more power plants. Don't build them near me. Don't build transmission lines."

On the oil side, the statements are, "Don't transport oil. Don't spill oil. Don't burn oil. Don't drill for it. Other than that, I want to drive my car." When I was trying to open up the Outer Continental Shelf for oil exploration off the coast of California in a series of public hearings, I kept hearing, "Don't put drilling rigs anywhere where we can see them." In the face of this kind of thinking, it's hard to imagine a change of attitude unless there's some enormous crisis.

Continued exodus

I see the integrated oil companies continuing to be distributors, refiners, and producers of some wells on a declining basis. As the wells decline in value, the majors will get out of them because they carry such heavy overheads. They have such large legal, executive, and communications staffs that it costs them a lot more money to manage a barrel of oil than it does a small independent. As the fields deplete, they will sooner or later say quit—like Texaco, which just got out of, what, 300 fields?

There's only one reason they did it. They thought they could use their money better someplace else. I don't know what the economics of these fields were. All I know is the Texaco management and board had to believe the long-term prospects from these fields are declining or provide inadequate margins. These fields can be sold right now for cash. Texaco can take the cash and put it someplace else and make more money. I think that is the long-term direction for the majors.

The larger independents will follow the majors. Their overheads are less, so they will operate longer in the U.S., but they'll be looking for opportunities to increase their returns by going overseas where the oil is more plentiful or the government restrictions are less, or both. Labor costs overseas may be less. The further down the chain you go, the less costly it is to operate.

A small mom/pop operator can operate a few wells and all he has to make is enough money to pay for his operating costs and to put some money in his pocket at the end of the day. He doesn't have the overhead, he doesn't pay lawyers, he doesn't pay anything. Relatively speaking, he hasn't got any overhead. So he can afford to operate long after the time others would quit.

I just think you're going to see a general exodus out of this country on oil. The big players will go overseas. I feel it will become extremely difficult to find money for exploration in this country. Domestically, it has to be seen as a high-risk investment—extremely high-risk investment—because you know that you're putting your money into looking for high-cost oil, which is vulnerable to a dramatic price decrease. Exactly the same thinking goes into solar, wind, ethanol—you name it.

The historic pattern in this country has been one of independents finding the oil and then joining together with majors to develop it. It is not going to be the pattern as much in the future, certainly not in this country. Overseas, the majors are the early players looking for the big—the giant—fields. I would expect initially that the majors and some large international independents—and of course foreign government companies—will be the ones to develop the big finds and build the big refineries, petrochemical plants, and the like.

Over time, if the international economic system will permit it, it's conceivable that you could get some large independents who might play much the same role internationally that they've played domestically. One of the reasons for that is that I think it's easier when you're no longer looking at the biggest—the Prudhoe Bay-type prospects—you've got a better chance that the smaller operations with some really knowledgeable and experienced people at the top can make decisions about where to go and what to do and be more efficient and have a better chance of wildcatting than a major. That has certainly been the history domestically.

The service sector is responsible for maintaining the infrastructure of the industry. I don't see how it can be revived without a major drilling program in the U.S., because that business has to follow the oil drilling business. Where people are exploring will determine where the service companies go. Service companies will take on a different role, domestically, and will prosper to the extent that they have processes that are cheaper for reworking old wells, or microbiological systems, or heat treating, or water flooding type things.

Secondary and tertiary recovery techniques may flourish here, because we've still got a tremendous amount of oil in the ground if we can find economical ways to get it out. So if somebody comes along with better ideas in that arena, he will still have a lot to do. There will be some reworking of wells that would produce at levels that are considered economically feasible. That level will vary depending on the world price.

Natural gas has a major and even expanding role to play, but I feel we're counting on it too heavily. Electric utilities have failed to survey the market correctly, and as they are planning very large system additions dependent upon long-term gas supplies, I think they're being misled.

I don't see a long-term gas supply that is economically available in ever-increasing volumes, 22 to 23 Tcf a year and rising. Maybe they can do that level for a long time with only a moderate rise in price. But I just don't believe that much greater quantities are there without more significant price increases. Maybe if you could open up all of the promising acreage in the Rocky Mountains and offshore and so forth, you could go longer. But without something like that, ever-increasing gas consumption without sharper price increases does not appear to me to be a long-term viable strategy.

Where we failed

Looking back to the 1980s, getting the message out regarding the importance of the domestic energy industry turned out to be impossible. I mean, it wasn't hard—it was impossible. I'll put it differently—we didn't accomplish it. We did not succeed. Whether it could have been done in a different way, you can always speculate. But I have yet to hear a persuasive presentation of what we should have done that we didn't do that would have been more likely to work.

The deck is and was stacked against responsible presentations regarding resource development and energy consumption. Sloganeers with critical or negative messages get at least equal and usually greater-than-equal treatment in the media. And the public is readily misled on that basis.

Compounding it is the fact that the industry speaks with multiple voices. At the very time we were fighting hardest to get ANWR open, I found out that two of the majors were running around Capitol Hill with their lobbyists, saying, "Don't open ANWR; we don't need it. Can't use it." That kind of a message is absolutely destructive.

So how does the industry speak? It doesn't. It speaks very badly. It always has and always will, I think, because it is made up of so many different voices. Take the majors that are tied into OPEC. They're absolutely hamstrung. One of the things that somebody needs to write a book about is the extent to which major OPEC participants retain most major law firms in America and have in their employ most of the major PR firms. When you think about it, Saudi, Kuwaiti, UAE, you name it, companies come into the United States and they retain the best.

So essentially all of the big players are probably in a position where they don't want to raise any alarm about what could be happening to the domestic energy industry. A major oil company that has billions of dollars invested in the Middle East is going to think twice about standing up for anything that would benefit the domestic oil industry in America to the disadvantage of OPEC. They can't afford to lose their concessions.

We've become extremely vulnerable as our reliance on imports has increased. Not to an interruption in supply, because truly unless the producers stopped producing so it simply wasn't available to the world, they can't stop oil from being bought by us. They're going to produce the oil. They're going to sell it through Amsterdam and maybe, as back in the late '70s, we'll be paying more money for it, but we'll still get it.

The long-term prospect is that, eventually, either of two things happens over the next ten and twenty years. The most likely is that the price will go up. The higher price will cause downward pressure on our economy but will excite investors overseas, and they will find additional giant fields which will bring the price back down. You can just see this playing through the world energy market again, because there are some enormous untapped resources of oil still to be found in the world.

As long as that's the case, if you find and develop them, there's no way, barring major breakthroughs in technology, that renewables and alternatives can compete with them. What that says is twenty or forty years from now, the world energy economy will still be basically petroleum based, and the United States will be increasingly vulnerable to economic swings because of upsets in world oil supply.

The alternative is a national energy policy which says we are no longer going to pretend the world oil market is either a free market or stable, and, therefore, we will provide some stability in order to foster domestic energy supplies. To do this will require a level of sophistication and commitment which would be unprecedented.

A New Vision

KENNETH L. LAY

chairman and CEO of Enron Corporation

Using innovative business practices has enabled Houston-based Enron Corporation to evolve into one of the largest energy companies in the world,
generating nearly $9 billion in revenues in 1994 with $12 billion in assets at
year-end.[1] The man largely responsible for building Enron into the world's
first natural gas major is chairman and CEO Kenneth L. Lay. A native of
Missouri, Lay graduated with honors in economics from the University of

1. Enron Corporation is engaged primarily in oil and gas exploration and production, liquid fuels extraction and marketing, natural gas transportation and marketing,
electricity generation and marketing, and financial services for the energy industry.

Missouri, where he also received a master's in economics and was elected to membership in Phi Beta Kappa. He later earned a Ph.D. in economics from the University of Houston.

Heavily recruited after graduation, he preferred a career in energy. "It really wasn't an industry you thought much about growing up in Missouri," Lay recalls. "But in the process of my interviews I became very fascinated by the energy business because of its importance to the economy, the country, and the world. I was very impressed with the capital-intensive nature of it, which, with my economics background, I found very interesting versus a more labor-intensive type of industry. It required a lot of long-range planning and interface with government and regulatory bodies, which also tended to fit into my economics training."

Lay began his career in 1965 as a corporate economist with Humble Oil & Refining Company (now Exxon Company, U.S.A.). He left to serve as a naval officer in the Pentagon and for several years worked in energy-related positions for the civilian side of the federal government in Washington. He was president of Continental Resources Company (formerly Florida Gas Company) before joining Houston's Transco Energy Company in 1982 as president, chief operating officer, and a director. In June 1984 he joined Houston Natural Gas as chairman and chief executive officer. Lay was named president and chief operating officer of Enron Corporation in July 1985 upon the merger of Houston Natural Gas and InterNorth Inc. He was elected chief executive officer in November 1985 and chairman in February 1986.

Lay was chairman of the Houston Host Committee for the 1992 Republican National Convention and cochairman of the 1990 Houston Economic Summit Host Committee. He is also former chairman of the Greater Houston Partnership and the University of Houston System Board of Regents. It becomes clear, when talking to him in his office on the fiftieth floor of the Enron Tower, that this man does not really move mountains: he has others move them for him.

I would like to describe the new natural gas industry in relation to the major-dominated oil industry model, explore some emerging themes in the increasingly competitive global energy marketplace, and examine some of Enron's strategies in a challenging environment.

The new natural gas industry

There are at least two overriding trends impacting the natural gas industry today. The first is that natural gas is increasingly becoming a global industry, unlike before when its member firms were either regional or national in scope. This globalization has resulted from the rapid growth of gas demand in virtually every region of the world, particularly for power generation. By just about anyone's forecast, total natural gas and liquefied natural gas demand is expected to grow twice as fast as oil consumption. This creates tremendous opportunities and challenges, particularly in the context of the worldwide privatization initiatives we are seeing. The major players in the natural gas business will increasingly become global companies, whether headquartered in the United States, the United Kingdom, Australia, or elsewhere.

This transformation, which is a whole new way that traditional gas companies have looked at themselves, is much like what the oil industry went through almost a century ago. The new breed of gas company will be able to offer all the different capabilities—from finance to construction to supply to marketing—to make a large, integrated infrastructure project a reality. We believe, in this regard, that Enron Corp. has achieved the status of becoming the world's first natural gas major.

The second concurrent factor driving the new natural gas industry is increased competition at home and abroad. It began here in the United States in the mid 1980s, when the monopoly franchise model began to be replaced with the "open access" competitive model. [Open access means that end users procure their gas from the most competitive source and turn to the traditional provider for transportation service only.] This has been accomplished for large end users such as electric utilities and industrials and is just reaching the residential and commercial level.

Adjunct to open access has been the formation of many new industry participants, particularly in natural gas marketing, that have facilitated competitive supply procurement for end users. We now have a national transportation grid and transparent prices for both the commodity and transportation in open-access markets, something that was unheard of in the industry a decade or two ago. The results have been phenomenal. Natural gas usage has expanded 20 percent, and real gas prices have dropped by one-third in the first decade of the open access era. Not surprisingly, economists are now using the open-access experience with gas as the model for restructuring the U.S. electric industry.

What has happened in the United States over the last decade has begun to occur in the United Kingdom and is starting to take root on the European continent. Other countries, like Argentina and Australia, are beginning to embrace the private, competitive model. As more and more of the international market moves toward competition, the whole industry will be forced to be market-driven, cost-driven, and flexible—all the things that a competitive industry has to be that a regulated industry does not have to be.

The oil majors and the new independents

The oil industry has some challenging times ahead of it. As the formerly regulated private industry, and in some cases state-owned gas companies, become more competitive and entrepreneurial, we could see the oil industry go in the opposite direction. Ironically, the very industry that was the epitome of entrepreneurship in the last century and for most of this century is beginning to look more and more like a regulated industry or monopoly. Part of the reason for this is their overall size. The oil majors are so enormous when you look at their cash flows, earnings, and employees. It is increasingly difficult for them to have the kind of entrepreneurship that they were built on and to be as responsive and risk-taking as they were in their prime. Downsizings or rightsizings have occurred with many, if not all, of the oil majors, but cultural rigidities have remained.

Corporate culture is very difficult to change. At one extreme there is an IBM that had to practically face financial disaster before major cultural changes were made. Given a bureaucratic culture, you typically keep making incremental decisions until a major financial crisis occurs or is about to occur. None of the oil majors have reached this bottom, but the tyranny of content or even confidence while the whole world is changing around you is there.

Many of the oil majors have big natural gas operations, but if you look at their assets, cash flows, and profits, 80 percent to 90 percent of everything they do is still oil related, and that's going to be a relatively slow-growing industry the next couple of decades, if the projections of oil demand are right. I don't want to paint too bleak a picture about the majors, however. Given their enormous resources, starting with their cash flows and all of their reserves and other assets, they will continue to grow and provide good returns for their shareholders. But they're going to have a very difficult time growing at the rate that the really good natural gas companies can grow or the really good independents can grow over the next several years.

Let's focus on the independents, where an interesting development has occurred. With virtually the whole world today deciding to liberalize their economies by going to capitalism, competition, and privatization, many smaller independents have sprung up around the world to compete with the oil majors. In markets like Argentina, for example, we are witnessing real growth among independents, which are entrepreneurial, agile, responsive, quick, and risk-taking, pretty much like the independents in this country historically have been and for the most part still are. This is creating a new genre of competition, which leaves the enormous capital-intensive, risky, and long-term projects, like Sakhalin Island or Kazakhstan, for the majors. It is the mega-projects that the majors are increasingly focusing on because of their size and because in many other markets, these little independents are moving in fast and taking opportunities away from them.

Independents are increasing their niche not only internationally, but in the U.S. upstream market. The oil majors have deemphasized their onshore operations and shed assets where they could not be competitive cost-wise. They are doing most of the deep drilling in the Gulf of Mexico and on some big structures onshore, but they're finding it increasingly difficult to compete in many basins with independent competitors.

The outlook in both the oil and gas industry for the independents is bright. The talented new independents from newly privatized economies are going to be one more thorn in the flesh of the oil majors, which historically could step in and get the big concessions and provide the capital. And certainly PDVSA, the Venezuelan oil company, and Pertamina, the Indonesian oil company, were more comfortable with them, since they were the private sector analogs to the big state-owned, monopoly-type organizations. But this is changing. Much of the new activity in South America, Asia, and the Middle East will be undertaken by medium-size or large independents.

Oil will obviously remain a very important component of the whole energy pie for at least the next century for three reasons: oil's piece is so big today, petroleum has a strong inherent advantage as transportation fuel, and there is so much supply around the world. But keep your eye on natural gas and the new independents.

Gas projections

Enron's most recent forecast estimates that world gas demand will reach 116 trillion cubic feet by 2010, which is also within the range of the forecasts by the International Energy Agency and the Department of Energy. What is striking is that natural gas demand will be growing at about twice the rate of oil demand during this fifteen-year period.[2] The world will need an additional 130 quadrillion BTUs of energy per year by the end of this period, and Enron expects 41 quads of this need to be satisfied by natural gas, 35 quads by oil, 30 quads by coal, and 24 quads by hydropower and other renewables. So although oil is by far the largest portion of the world's energy pie today, fewer quads of BTUs will be added to this pie over the next fifteen years than quads of natural gas.

As optimistic as this forecast is for gas compared to the other fossil fuels, gas demand will be much more pronounced if, in fact, the scientific uncertainty of global warming becomes a more scientific certainty. If you look at all the forecasts, including our own, carbon dioxide emissions from energy consumption are expected to increase by about one third between now and 2010. If the industrialized world is committed to reducing carbon emissions back to 1990 levels by the year 2000, which is what these countries are saying, some dramatic changes will have to occur. Consequently, the use of coal and oil may face even further political pressure.

I am not as enthusiastic about natural gas vehicles as some people are. I think this market will grow in the U.S. and may eventually be as

---

2. Enron's 1995 Energy Outlook estimates world natural gas demand growth to grow 3 percent per year over the 1994–2010 time frame, compared with total energy growth at 2.2 percent. Oil and coal growth are 1.6 percent and 1.9 percent per year, respectively. Enron estimates the recoverable natural gas resource base to be 14,024 Tcf as of January 1, 1995. This includes 5,040 Tcf of proved reserves.

significant as it is in other countries such as Argentina and Russia, but it will still be a very small piece of the total U.S. or worldwide energy picture.

The inherent advantages of natural gas, outside of its chemical feed-stock uses, are still going to be in its stationary uses—power plants, industrial boilers, and buildings. I believe power generation will over-whelm all the other sectors as far as growth, and that's mainly because of the advances in natural gas combined-cycle power plant technology.[3]

Compensating for success

Enron's whole workplace environment and strategy reflect the fact that the energy business is becoming more and more of a knowledge indus-try. It is not that Enron does not have a lot of hardware, but increas-ingly it is going to become more of an information industry and less of a hard asset-driven industry.

An important part of our corporate culture is individualized compen-sation in each of our business activities, many of which are new areas for us. We have different compensation systems in Enron Oil & Gas [EOG] than in our pipeline business and different compensation sys-tems in our marketing and finance business than in our international development business, for example.

It would be a major change for the oil majors to significantly deviate from their hierarchical, one-size-fits-all compensation system that has been part of their culture. It really takes a unique compensation struc-ture to compete against the investment bankers or the entrepreneurs that are developing projects and living in international destinations for months at a time, because they realize that if they put a successful pro-ject together, they can hit a grand-slam home run financially for their family.

I have been very satisfied with the results of Enron's compensation program to date. We have lost very few of our key developers, traders, or executives that we wanted to keep. We base a large percentage of our

---

3. Enron estimates that natural gas demand for power generation in the U.S. will rise from 4.3 Tcf in 1994 to 10.1 Tcf in 2010. In comparison, residential use is esti-mated to rise from 4.9 Tcf to 5.2 Tcf. The U.S. consumed about 21 Tcf in 1994, a figure expected to rise to 29 Tcf by 2010, says Enron.

compensation on performance, particularly long-term performance. Obviously, this requires the right structure so that people stay around three, four, or five years to get the full payout.

There are big payouts. Forrest Hoglund, chairman, president, and CEO of Enron Oil & Gas, topped a local list of the most highly compensated executives in 1994, with a remuneration of around $19 million, primarily through stock option exercises. There's not a CEO of any major oil company in the country that made nearly that much. But that compensation was based almost entirely on the performance of EOG. He has quadrupled the market value of that company in about six or seven years in a very tough market, and he is presiding over a company making 15 percent to 16 percent return on equity. That's very good in the oil and gas business today, when the best independents are making closer to a 5 percent return. He obviously created a lot of wealth for himself, but he has also created a lot more wealth for his shareholders, including Enron Corp. But the philosophy to let this happen is just not there in most major oil companies.

Hedging for success

Enron's entrepreneurship has both risk-taking and risk-mitigating components. We try to hedge the prices of most of our natural gas, our more modest oil production, and the interest rates we pay. We have a financial plan, and we try to lock in the contract volumes and prices to ensure we meet it. We then concentrate on the things we have control over, such as our costs, volumes, and new product development. That is the reason we have had six or seven very strong growth years back to back, even in difficult markets.

Many other companies refuse to hedge their oil and gas prices. They prefer to speculate that there may be a big price spike soon. The majors are often slow to hedge for different reasons than the independents. Their problem is that the decision makers deciding to hedge or not are down at lower levels, and there is very little incentive, and probably a strong disincentive, for possibly guessing wrong. For example, you will be criticized if you lock in prices in a rising market, but will probably not be rewarded if you lock in prices when they drop. The reward will most certainly be less than the penalty without a hedging strategy in place.

It seems as if many independents are convinced that around that next corner—that next winter or that next oil embargo—there is going to be a big price surge to allow them to even up with the hard times already experienced. They will be right occasionally, just like they were during Desert Shield/Desert Storm. But the long periods in between these occasional and unpredictable spikes require that the physical business be run more like a financial business.

We are also applying our risk-trading and financial tools in international markets—Europe and South America initially. We think over time you will see those markets evolve through competition and deregulation much the way the U.S. market has with natural gas and electricity.

Selling know-how in a demand-driven market

Enron is well positioned in the global energy market. First of all, we believe we have picked the right businesses at the right time. We are an integrated natural gas company with the capacity to participate in any aspect of the business. Secondly, we have earned our spurs in the most highly competitive energy market in the world. Having become the U.S. leader in our industry has given us a big edge when we enter new markets around the world.

The main thing a company like ours sells internationally is the capability to put together large, complicated infrastructure projects. That basically comes down to knowledge, information, entrepreneurial skills, and being able to assemble all the appropriate financial and purchase agreements, the supplies, the sales, the turnkey contracts, and the management to build and operate the infrastructure projects.

Putting these projects together is a difficult task. Actually supplying natural gas or oil to them is not that difficult, but it is different than the traditional supply-driven approach.

The international oil industry historically has been a supply-driven business. Companies in this industry have tried to control the upstream supply and felt if they did that at the right cost, then they did not have to worry as much about downstream profits. There are cases where they've actually lost money in refining or marketing to move production from their upstream business. In contrast, today's natural gas market is mainly demand-driven, where knowledge, market responsiveness, and entrepreneurship are the source of competitive advantage.

We expect to see some restructuring of the world LNG markets, which the majors and producer-country governments developed as a supply-driven business. Traditionally, they work up the costs, build the project, and then sell LNG on a cost-plus basis. Because of this approach, the LNG industry is currently having problems with proposed new projects. The LNG industry has to determine what the market will pay and then figure out if it can get the cost of LNG down sufficiently to satisfy the market.

Regulatory issues

In a global market, a nation's capital-cost recovery system has to be competitive, which brings us to tax policy. If the tax burdens on oil or gas increase enough, the industry could be driven out of the country or certainly made noncompetitive in the international market. The U.S. must have an internationally competitive capital-cost recovery system. Beyond that, for the most part, governments should stay out of the energy business.

The U.S. capital-cost recovery system today, particularly with energy, is not optimal. Throughout the '70s and part of the '80s, Congress kept chipping away at the industry, because energy was out of favor in Washington. If there is one item in the current tax code that needs to be looked at from the standpoint of the energy industry and many other industries, it would be the alternative minimum tax. AMT is an extremely onerous tax on oil and gas producers that has hammered the industry in the last few years, particularly the independents. It is not a very rational tax. When oil and gas prices are low, producers are forced to pay an even higher percentage of the tax on their lower revenue and cash flow. The AMT penalizes everyone, even the good performers.

Regarding domestic drilling, I support more access to our prospective areas. I think in many cases we have blocked access to promising areas more on emotion than on any type of environmental science.

Having said this, Enron, and I personally, would be concerned if the country started to retrench on its core environmental goals. As a society, particularly given the potential risk of scary outcomes suggested by some of the global warming scenarios, it would be a real mistake to back down on our high environmental standards, particularly for water and air.

We need to look very hard at the means used to set environmental policy today. There are many cases where we could achieve our goals much more efficiently through more market-driven solutions, such as giving flexibility to companies or industries to achieve or even over-achieve the goals. For example, we had a PCB disposal issue a few years ago, as many pipeline companies did. We started to look at the various disposal options and discovered that the EPA in one region had approved a method of encasing PCBs in place which ensured they would never again be exposed to any water or air. This method was a fraction of the cost of trucking them to a dump site in a remote location. But on the technicality that we were in a different region than where the EPA had approved the encasement procedure, we never got approval, although we went all the way to the EPA administrator in Washington in our attempt. That is not smart. There is no reason we should be forced to spend more money and choose an inferior disposal option than one which has been proven elsewhere—that is bad economic and environmental policy.

In many cases, if you let the market decide how to achieve particular goals at the least cost, natural gas is a big winner. We have the situation now in the U.S. where if you repower an old coal plant to continue to burn coal, you emit more air and water pollutants than if you repowered the same plant for gas. Let the market choose between coal or gas. Absent perverse incentives, such as higher capital costs that yield greater regulated returns, the market will choose natural gas with today's options.

Restructuring the U.S. electricity industry

We are increasingly going to see commoditization of natural gas versus electricity, and vice versa. As you continue to deregulate these markets, the national energy market will become more competitive and rational, replacing markets that were highly rigid and structured by regulators. Keep in mind that consumers do not care if it's electricity, gas, oil, or some other fuel. Consumers are only interested in satisfying their energy needs at the least cost.

I expect to see a lot of consolidation in the electric industry. For a large industry, electric utilities are one of the most fragmented in the world. This reflects its regulatory past and does not make sense from an economic efficiency or consumer perspective. These consolidations will bring forth new leadership, as I will discuss later.

State-level public utility regulation is a complication for where the electric industry needs to go. Like we have seen with natural gas and other deregulated industries, the regulators will eventually follow the markets. Once a few states provide for competition, industries in those states will enjoy lower electricity rates and energy costs. This will create enormous pressure on the hold-out states as companies move plants or reload plants in the low-cost jurisdictions.

This competitive dynamic is ready to go. The regulatory reform to make it happen is falling into place. The Energy Policy Act of 1992 legalized competitive procurement of electricity at the wholesale level, and FERC is moving forward to administratively provide nondiscriminatory transmission access for third parties. Wholesale electricity in this country is about a $90 billion a year market, which is three times the size of the gas wholesale market and about one and a half times the size of the oil wholesale market.

The next step in the electric restructuring is retail wheeling,[4] which we don't really have yet in the gas business. But it is starting to occur in the gas industry, and it might happen even faster in the electric industry.

The "POOLCO" model for electricity spot markets, however, is a threat to retail wheeling, also called customer choice. POOLCO is an invented market where regulators force all transactions to be centrally dispatched from one regional pool. Generators must sell into the pool;

---

4. Retail wheeling refers to the transmission of electricity generated by a non-utility over the lines of a utility. Utility companies complain that IPPs (independent power producers such as Enron) can offer an attractive price because they don't have the fixed costs of a utility.

end users must buy through the pool. The free-market model, in contrast, allows bilateral and multilateral contracting according to the market's infinite variety, while retaining central grid control for physical distribution to ensure system integrity. End users want a free market in electricity, as do marketers such as Enron. Utilities with high-cost generation who desire to maintain monopoly control, not surprisingly, are lobbying hard for POOLCO.

We feel strongly that POOLCO could be the next regulatory tar baby. Markets are not manufactured. Markets evolve. Competition in the gas industry evolved in ways that no one could have predicted or planned, and the experience with electricity will not be different. It is ironic that with the current intellectual and political momentum toward markets, POOLCO could attract as much interest as it has.

Markets are strong things. Certainly nobody in the electric industry had any idea that the Public Utility Regulatory Policies Act of 1978, or PURPA for short, would set in motion something that will eventually cause that whole industry to be restructured along competitive lines. The Natural Gas Policy Act did the same thing. Few people in 1978 thought that if some prices for wellhead gas were deregulated, it would be only a matter of time before all producer prices had to be deregulated and whole new regulatory regimes would emerge as a result. Competition simply cannot be quarantined, as one participant in the California electricity hearings stated. Yet we see attempts to do so in the electric industry today.

Changing leadership for the competitive era

The new leadership in the evolving gas and electric industries will be increasingly market-oriented, information-oriented, and probably less engineering- and lawyer-oriented. It all gets back to globalization and competition. Globalization means that you need chief executive officers and others who have the capability to understand governments and politics in multiple countries. Competition means you need executives who have a very strong market and financial orientation. I predict that the future leaders across the energy industry will be more generalists, particularly from financial or marketing backgrounds, compared to the engineering backgrounds of before.

I think you will also see natural gas people going into other industries. We're seeing some of that now. One of our former senior executives became a CEO of an electric utility. Other Enron executives have been approached by large electric utilities, because they want to get some talent that has gone through the natural gas transition and has some idea of what may be coming down the road. The natural gas industry has had a fairly dramatic turnover of senior management over the last ten years—I think you'll see the same thing at the electric utilities. Some of it will be evolutionary from within, obviously, but I expect you'll also see more from the outside, too.

Problems of high-cost U.S. power

Integrating the nation's high-cost electric generation into a competitive market will be a major transition. There will be cost sharing between ratepayers and shareholders, as was the case with the uneconomic natural gas contracts held by interstate pipelines a few years ago.

Nuclear power will be the biggest problem in the transition to competition, not only because of its high all-inclusive cost, but its high marginal costs as well. The life cycle characteristics of nuclear plants, and to a lesser extent coal plants, require very expensive repair or replacement due to aging degradation. Low-level radioactive waste storage and disposal facilities for nuclear plants are another cost that is not reflected in current costs and prices for nuclear-generated electricity.

A return of commercial nuclear power will be slow. I will not say that a reemergence of the nuclear industry will never occur, particularly if global warming becomes the big issue. But nuclear power is a long way from being competitive with gas-fired combined-cycle plants that are getting more efficient all the time.

We think that when you really get a competitive generation market and start dispatching based on market economics rather than regulatory incentives, you will see many nuclear plants shut down. The problem of nuclear is much more than just safety risk; it is relative economics.

Obviously, there are some very efficient, cheaper nuclear plants just like there are some extremely inefficient ones. But the average facility, just based on the amount of money you have to keep pouring into it to keep it operating, makes a shift to new natural gas combined-cycle plant technologies increasingly cost-effective.

Conclusion

We have been through some very exciting and challenging times in the last decade that have set the stage for more of the same—competition and globalization with natural gas at the forefront. The next decade will see the U.S. electricity industry competitively reshaped and a boom in the international power market. Here at Enron we have our work cut out for us. We achieved our first vision of being the world's first natural gas major. Our new vision as a global energy company is to become "the world's leading energy company, creating innovative and efficient energy solutions for growing economies and a better environment worldwide." Perhaps a future edition of your book can see how well we did on this vision.

Paul M. Anderson

president and CEO of Panhandle Eastern Corporation

Paul M. Anderson is an executive who can't resist challenges. He has been at the forefront of nearly all of them during the deregulation of the natural gas industry in the past fifteen years. Today he is president and chief executive officer of Panhandle Eastern Corporation, which operates one of the nation's largest interstate pipeline networks and is also one of the most aggressive gas gatherers and marketers.

Anderson, a native of Washington state, holds a B.S. degree in mechanical engineering from the University of Washington and an MBA from Stanford Graduate School of Business. He worked eight years in management for Ford Motor Company before joining Texas Eastern Corporation in 1977. He helped restructure Texas Eastern during the 1980s and was its lead negotiator when it merged with Panhandle in 1989.

In 1990, Anderson was working for Inland Steel Industries in Chicago when Dennis Hendrix, his former boss at Texas Eastern who was about to become chairman of Panhandle, asked him to become a member of the turnaround team at troubled Panhandle. Several months later, Anderson returned to Houston, where he helped devise the strategy that has led to Panhandle's well-documented resurgence. Among his tasks was resolving unsolved rate cases and litigation dating back ten years and restoring the company's relationships with its customers. In 1994, he proposed a merger with Associated Natural Gas Corporation of Denver, a major independent company, and was responsible for completing the $830 million deal that vaulted Panhandle into the upper echelon of the industry.

Analysts and consultants say Anderson is the prototype of the new executive in the gas industry, energetic and expertly able to envision the position of gas in the total sum of the energy picture. He also foresees a day in the near future when energy supply and demand will be handled with software, as the industry becomes less dependent on its standard hardware assets. In this conversation, he looked toward that future with an enlightened and stimulating "we can do it" approach.

If you look at the whole industry, probably the biggest challenge I see is to have the industry come to grips with the commoditization of energy in total. There has been a lot of talk about the commoditization of natural gas or the commoditization of oil—the components of energy.

But what I see happening between now and the year 2000 is that energy per se will get commoditized, and there will be interchangeability between various forms of energy. Heretofore, we've had the segments somewhat protected from each other. There's been a gas industry, oil industry, electric industry, coal industry. I see the barriers between those segments blurring and getting broken down, so that basically all of energy gets commoditized.

What that means is that energy is no longer special the way people have viewed it in the past. The basis for competition will have more dimensions, because you have to worry about not only can you find oil and gas competitively with another oil and gas producer—you have to worry about whether oil and gas will be competitive to start with in terms of your price forecast. Whether interchangeability of electricity or "coal by wire" will be competing with gas, for instance.

We have a lot of fuel switching in the Northeast that allows end users to go between gas and oil. That has affected the ultimate price cap on gas.

Because we're now going to have fuel switchers going from electricity to gas, they can indirectly go from coal to gas or oil. So all the various forms will start competing with each other on a much more commoditized basis. That means you're going to do business totally differently. The industry is going to get much more efficient.

There have been tremendous inefficiencies built in. The industry was developed over the years based on inefficiencies. As those inefficiencies go out, it's going to continually amaze people how much gas can be produced with so little drilling. It's going to continually amaze people why the reserve/production ratio can go down and we still don't see prices go up. It's going to continually amaze them that deliverability relative to demand will come closer and closer together and we don't see more price spikes.

And it's because of all these inefficiencies that are going to be wrung out of the system. Ten years ago, people said if the rig count ever drops below 1,000, we're going to have a disaster on our hands. Today people say if the rig count ever hits 1,000, we're going to have a wonderful day. I just saw an American Gas Association report[1] that said the replacement of reserves was well over 100 percent for 1994 in the U.S., and the number of reserves added in 1994 was higher than any year since either '81 or '82, when we had a rig count of over 4,000. There were very few wells drilled relative to what we had in the early '80s. What we're seeing are deeper wells, more efficient drilling. Obviously, identification of prospects has come a long way. It said actual additions of reserves in 1994 may surpass the 21.5 trillion added during the drilling boom of 1981.

---

1. The report released May 12, 1995, by the A.G.A. Policy & Analysis Group said natural gas reserve additions in the United States are expected to be between 105 percent and 137 percent of production for 1994, based on data from the top thirty reserve holders of natural gas.

The amount of efficiency that's gone into the system is tremendous. The restructuring of the pipeline industry has freed up a tremendous amount of transportation capacity simply by removing regulatory barriers, and it hasn't required that we add more pipe in the ground. The advent of storage has caused 365-day utilization of the grid, instead of peak-day utilization the way it was in the past. We built a pipeline from Texas to New York to handle peak-day requirements and expected deliverability in the Gulf of Mexico to be able to handle peak-day requirements in New York. Now with storage in New York, we still have a 365-day capacity based on peak day. We can fill that storage in the off days in the summer months and increase the effective capacity of the grid many fold.

I see the industry trying to adjust to this commoditization and stay ahead of the efficiency curve. A good analogy is to look at what's happened in computers, where somebody comes out with a faster chip and they say, "We'll never need it any faster than that." And then somebody comes out with an even faster one. Somebody figures out a way to use the speed of the last chip and actually stretch its limits and go to the next phase.

I see the energy industry going through the same thing. They're going to say, "We're just as tight as we can be on reserves-to-production." Then they're going to come up with the next increment of just-in-time planning through the system or fuel switching at the end-use that doesn't require as much margin because if you bump up against the limits of gas, you can switch to fuel oil. If you bump up against the limits of fuel oil, you can shut off electric generation from gas and bring gas in from another region that's generated by coal. That's where I see the big challenge.

Software versus hardware

It's going to be a much more software-oriented industry than a hardware-oriented industry. I believe that independent producers are going to become more like farmers and that they're going to have to hedge their production. When a farmer plants his field and hedges the corn crop going forward, he says, "I can't afford to bet on weather and what's going to move the price of corn twelve months from now. So I'll just hedge it, and depend on being an efficient farmer."

I think a lot of producers are going to have to do the same thing, because they won't be able to afford the swings, and the swings will be down as well as up. In the late '70s, early '80s, you had the beautiful situation where inflation covered a multitude of sins, and any capital investment looked good in that environment because you got immediate tax benefits. Your cash flow pattern versus your earnings pattern made you look smart no matter what. A lot of people were able to make marginal investments in that environment and they turned out very well. But if you're in a low-inflation environment and you don't have much volatility, and the volatility down is as much as the volatility up, and there's no real growth, you might have to hedge your prices going forward.

I believe this will evolve naturally in that those who still remember a 4,500-rig count are starting to retire slowly but surely. If you look at the new generation that's coming in, they don't remember the good old days. I feel that a good example of a forward-thinking company is Louis Dreyfuss Energy, a company that specializes in risk management products through physical contracts and financial instruments. They went from a trading company and getting involved in futures markets and playing one market against the other to saying, "Once we've got this software in place we can afford to buy some production over here because we know we've got a home for it. We can sell it forward."

They view the production of oil and gas as part of an overall managed financial program, and while they want to be efficient and effective operators, the value they derive has to do with managing the financial aspects of the business as well as the physical. To me, that's the next generation going forward. And, it will naturally evolve that way because you're going to get fresh blood into the system here.

I'm sure anybody who is inefficient will resist. Whenever you have a system that's evolving toward efficiency, the least-efficient operator wants to slow down the change as much as possible. We saw that in the pipeline industry as we went through the restructuring under Order 636. There were some companies that jumped ahead and embraced it, saying, "I'm going to go there first and I'm going to set the terms under which we do business in the future."

There were others that just dragged their heels and said, "We'll see you in court. We'll fight this tooth and nail," because they didn't want to have to compete on that faster track. I think you'll see the same thing in the electrics. In fact, you're already seeing that. You're seeing a couple of companies get out ahead of the curve and actually start merging with each other and trying to unbundle.

The gas transmission companies have already gone through the first phase of change of their business. Having done so, I believe their managements at least appreciate some of the aspects of the change: how fast it's likely to come, order of magnitude, and how to respond to it. So maybe psychologically they're set up for it. I'm not sure whether or not they're technically set up for it—electricity is a whole different ball game—of course, oil and gas is still a different ball game.

If there were one thing that I would say would characterize the industry going forward, it's that it is going to be biased toward software as opposed to hardware. It's going to be knowledge-based as opposed to physically based. The winners are going to be the ones that can play in several dimensions, including taking advantage of relative values of energy in terms of location, time, and form. In other words, knowing that there is a value to be arbitraged between coal in the Midwest in the summer versus electricity in the Northeast in the winter, and how you arbitrage that difference.

You do it by exchanging electric-generated power in the Midwest for gas-generated power, and store the gas someplace so you can deliver that in the winter to generate electricity then or to shut off electric generation and therefore free up gas. To be able to deal in those dimensions as opposed to simply linearly thinking, "I produce gas in the Gulf Coast. Somebody needs it in New York. I'll build a pipeline and get it from here to there." The least reward will go to that linear thinker who just thinks in terms of produce and deliver energy to an end user. The biggest reward is going to be the one who can arbitrage between the time, form, and location—differences in value.

You really won't have to own the properties to be involved. For instance, we might end up dealing in coal futures and arbitraging those against electric futures, against gas futures. But I don't want to own a coal mine. Now, I'm not saying we wouldn't. I think that some hardware is important to support a software system, but that you first figure out where you want to be on the software side, and then you decide what are the critical pieces of hardware you need.

I doubt that I will need a coal mine. I might need a cogen plant someplace that can fuel switch between gas and oil. That might be a critical piece of hardware because that acts as instantaneous storage, if you will. It is a way to exchange one form of energy for another, or act almost like a storage field because you can free up a gas supply simply by switching fuels. I would choose assets to support the software which would span various forms of energy. But I don't want to get into production of all forms of energy.

The gas age

My view might be a little bit different than the official industry view. The only places where it might have a home but it's not being used, it's limited by the distribution system as opposed to the technology. My view of the gas age is that gas will be used where it makes the most sense and where economics drive it. I think that most of the more exotic uses of natural gas are stretches. If it has a use, it's probably already there. For instance, stationary power plants. To me, there's nothing that makes more sense than generating electricity with gas.

If I were looking for incremental uses for gas it would be industrial and commercial installations, home heating, and electric generation. It would not be natural gas vehicles. That isn't a natural form. I think you're ahead to back liquids out of electric generation and put those liquids into transportation as opposed to trying to find new uses for gas in transportation. So I would see gas becoming a base-load fuel. It is by far the best base-load fuel for stationary users of fuel. It's basically coal with two or three degrees of freedom that make it more effective than coal.

Coal is a good base-load fuel for just plain electric generation, but gas you can distribute to homes. Coal is pretty tough to do that with. Gas is cleaner burning. You don't have all the issues of clean up. You don't have the ozone non-attainment areas to deal with because gas is environmentally a step ahead of everything else. It's a much more flexible base-load fuel. But it is a base-load fuel because it's not set up to store as easily as liquids. So your second level of fuel is a liquid, which you can store a bit more easily.

Now we're reaching the point with gas that we've got enough storage so that we're removing the seasonality; so it can be a base-load fuel without having the peak-day capacity problems that we used to have. One of the problems before was that if you used it as a base-load fuel and you wanted to use it in the heating season, it was too expensive to bring that next increment of capacity on board. But now we've got so much storage that we basically use the grid 365 days a year, eliminating that problem.

Low prices versus prosperity

The prices today are where the prices are and will be. I don't view to-day's prices as being abnormally low, even though a lot of people are talking about how low they are. I think they're probably about where they should be. They're just low relative to the aberration in gas prices back in the early '80s.

It's very hard to imagine that we're adding reserves at the rate we are because people are so hung up on the rig count. But what a lot of people don't appreciate is that you add reserves in ways other than just drilling wells. For instance, deliverability responds almost immediately to a rise in prices. The reason is that people can afford to put more compression in the field and lower field pressures, so wells produce at higher rates.

So if tomorrow you tell me the price of gas is going to be two bucks, I'll guarantee deliverability will go up. That's because every producer out there will crank up wellhead compression or he'll beat on the gatherer to lower field pressures and deliverability will go up. As time goes on beyond tomorrow, they'll go in and clean up old wells and re-complete wells and perforate another zone. None of that takes another drilling rig. It might take a workover rig, but it doesn't take drilling an-other well. So there's a lot you can do to add reserves without actually going out and wildcatting.

I don't feel that we're particularly vulnerable [to international pressures], even with a high dependence on liquids, because the sources of supply are becoming more diverse. As we spread out around the world and get more production from South America and the Far East and places other than the Middle East, I think we're less vulnerable to start with. As we get more interchangeability of fuels, it will definitely decrease our vulnerability because, to the extent that energy is energy and you shut off oil and you can replace it with coal or with natural gas, that obviously diminishes your dependence on the oil.

Pricing is getting much more predictable, much less volatile. We'll see dislocations and it will tend to occur when, for instance, a major line goes down because of a hurricane or we get a period of cold weather—we'll see little spikes. We're not making business decisions based on real growth in prices with anything above 1 percent. The long-term winners will be those who took the volatility out in planning their strategy.

Consolidation of pipelines

I would say that the acceleration of consolidation is going to be heaviest even outside the big interstates, because the interstates are pretty consolidated already. But as I see the industry going forward, basically the pipelines have had their franchises taken away. It used to be that they had a market area that was theirs and a supply area that was theirs, and they were the only route from here to there which nobody else got into.

The restructuring has broken down all the barriers so they don't have franchises any more. Now a landlocked pipeline that doesn't have access to multiple sources of supply and multiple markets has a very difficult time competing, because all it's selling is basic transportation. It can only leverage its hardware. It can't leverage its software of arbitrage between markets and supply basins.

I see the more landlocked of the pipelines, or the ones with less degrees of freedom, having to align themselves with somebody who has more. I think some of them are pretty obvious. There are some that go from point A to point B and that's about all they do. Those pipelines will align themselves with somebody that's got more of a web out there.

If they aren't aligned with a larger system, they'll tend to get put in a position where they have to discount their services because there'll be routes around them. The larger the grid that you're associated with, the better off you are as far as being able to have a reasonable rate structure and not get discounted to death, if you will.

So I see that even among the FERC-regulated pipelines there'll be some consolidation, and in the unregulated pipelines I see a lot of shifting around of asset use and ownership. We have a lot of depreciated gas lines that could better be used carrying liquids or oil. We're looking at a conversion right now on one of our Trunkline lines to crude oil. We have liquids lines that might be better carrying gas. We have gathering systems built by the majors because they were developing a field and the production of that field has gone down. There may be only 20 percent of the flow through that gathering system which is their own production.

They really don't want to operate a gathering system. Somebody else could operate that better and more efficiently. So a change in ownership there makes sense. I see a lot of shifting of assets. All the assets that we have in the country were put together under a different set of economic drivers than we have today. All the players have to sit back and ask, "What's driving it now? Does it make sense for me to own this? Or, what is out there that I can now do more efficiently?"

For instance, we are getting into gathering. And we're trying to buy gathering systems from producers, because we think that by putting gathering systems together and aggregating production from a number of producers, we can do a better job of running those systems and extracting value out of them than a major who's primarily focused on production. The majors, I would hope, are looking at it and saying, "I'm in the business of producing oil and gas. I'm not in the business of worrying about a gathering system, field compression, and all this other stuff. That's not my bag. I'd rather sell that to somebody and contract with them to just get my gas to market."

I think it will be harder to exist as a smaller anything in the marketplace going forward, because the barriers have broken down so much that it's going to be a lot like farming. What happened to the small family farm? Other than for emotional reasons, you really don't have them anymore. I think we're going to see the same thing. Any small entity, be it pipeline, producer, processor, any part of the chain up to the end user that is small, will have to be very specialized in what they do or they won't be able to compete.

Environmental issues

They play a major role in any company's planning. They offer opportunities, they offer constraints, and they offer problems. For instance, you might not think of it, but we have a major drive this year to meet the Clean Air Standards because we have more than a hundred compressor stations around the country and a lot of these are in nonattainment areas. If you're in a nonattainment area, you have to come up with a NOX-reduction plan.

So we're putting a lot of effort into NOX reduction at these compressor station sites. On the other hand, the fact that nonattainment areas are trying to reduce NOX produces opportunities for you to serve markets that were served by something else that was a dirtier fuel. People want to burn natural gas where they didn't before. So it's a market opportunity at the same time.

When you build a pipeline now, environmental permitting can take six months or so. It can be the biggest obstacle and the biggest cost. We estimate that about 20 percent of the cost of construction is now associated with permitting and environmental. So it's a big issue, but I don't think it's particularly unique to us.

LARRY W. BICKLE

chairman and CEO of Tejas Power Corporation

Houston-based Tejas Power Corporation is a growing gas gatherer and marketer that has astutely exploited the changes in the natural gas market caused by Order 636. The company was founded by chairman Larry Bickle and president John Strom in 1984 as a gas gatherer.

Bickle served on the technical staff of the Sandia National Laboratories, the government think tank that was responsible for developing the salt caverns in which the Strategic Petroleum Reserve is stored. He eventually applied that public-domain information to gas storage and bought three sites that were good salt domes with access to multiple pipelines. With deregulation on its way, Bickle went after a market niche and built a company that successfully fulfilled the needs of a competitive gas market.

As Tejas waited for the right time to develop its salt storage caverns, it sold a 25 percent share to Gaz de France, the French-owned gas company that is a world leader in salt storage technology. Today, Tejas is a leader in the development of the market center or "hub" concept, in which gas comes from wherever it is available to hubs and is then "parked" in salt storage until it is sent out to wherever it is needed. The salt storage provides the minute-to-minute balancing between the connecting pipelines.

Bickle holds a B.S. degree and an M.S. in mechanical engineering from the University of Texas. He also has a Ph.D. in mechanical engineering from the University of New Mexico. As part of a changing management style particularly seen in the natural gas industry, Tejas employees all own a piece of the company, giving them a vested interest in its well-being.

In this interview, conducted in March 1995, Bickle foresees the need for the natural gas industry to soundly restructure itself if gas is to become the preferred fuel of the future.

At the broadest level, it's clear that international competition is causing a realignment or readjustment of U.S. energy policy and practices. Specifically, as our industries go out and compete in international commerce, they can't afford to have subsidies to the energy industry embedded in the cost of their product.

If you think about an aluminum manufacturer, about 90 percent of the cost of aluminum is electricity. Obviously, the seller of aluminum can't afford to subsidize the other rate payers of the electric company. The aluminum manufacturer's customers don't care about the financial strength of Houston Lighting & Power. All they care about is the cost of aluminum in Kuala Lumpur, for example.

It's clear that the driving force of deregulation, particularly in the gas and the electric industry, is really about driving down the cost of energy as a component in the products that are sold in international commerce. I don't see that trend going away. If anything, I see that trend accelerating in the future. There is no question that deregulation has caused a lot of trauma, dislocation, and upset. The numbers, which I have not verified independently but which have been quoted in the *Wall Street Journal*, are that in the '80s, the energy industry lost a half million jobs. That trend will probably continue, as there is more and more pressure to be efficient.

The gas era

I do not share, for example [Enron Corporation chairman] Ken Lay's enthusiasm about natural gas as a fuel of the future. I think natural gas is positioned to possibly be the fuel of the future. But it is also positioned to decline as well. I think the natural gas industry is at the crossroads for the next three to five years. What all of us in the industry do during this short time period will determine whether natural gas reengineers, restructures, and revitalizes itself as the U.S. automobile industry and computer chip manufacturers have done, or whether the gas industry declines to become an economic backwater like the railroads did.

Need to reengineer

Prior to deregulation, there were a bunch of individual pipeline systems, each connected to its own dedicated supply and market. While they interconnected and occasionally exchanged gas during times of crisis, they really didn't work together to optimize the transportation of gas as a whole. Each was managed by competent, qualified people, and each optimized its own individual subsystem in a very quantitative, professional way.

But there is a theorem in systems engineering that says you can't optimize a large, complex system by optimizing the individual subsystems. What deregulation really did was provide an opportunity to switch from optimizing individual pipelines to optimizing the whole transportation grid for moving gas from the production area to the market area.

Optimizing the transportation grid really translates into the simple concept of letting gas go from where it is available to where it is needed. I would argue that the gas bubble—the gas bubble that's become a gas sausage that keeps on going—is primarily the result of the changes that occurred in the transportation system. As existing wells are used more efficiently, fewer wells are needed.

I don't want to come across as negative, but I'm concerned that the efficiencies we've won by integrating the individual pipelines into a grid will be lost. Unless regulators continue to create a fertile environment for new, enhanced efficiency devices to come into the market, they're going to be suppressed by the old oligopoly. And the financial community will help the old oligopoly.

The pipelines are buying up all the deregulated companies that sprang up. As this occurs, I see a real danger of a return to the old oligopoly—different faces, different names—but, nevertheless, loss of the technical upside that was created by the original deregulation. I think it leads you into a stagnant gas industry.

Restructuring deliverability

There is one very important broad concept that needs to be debated, analyzed, and either accepted or rejected by everyone in the gas industry. The issue is whether the gas industry intends to optimize gas delivery as an overall system from the wellhead through the burner tip, including production, transportation, and distribution. Or, do we have a greater obligation, because of concepts of stranded assets and so forth, to reject optimizing the overall system and instead focus on optimizing the historical individual subsystems?

The existing transportation system is much less efficient than what we could have in the future. There are two basic ways that gas can be delivered. In the past, gas was dedicated to a pipeline and went along that pipeline until it got to the specific pipeline markets. The markets only had access to the gas that was tied to that pipeline. More recently, the concept of market hubs was introduced. A market hub is a place where a bunch of pipes come together so that gas can come in on any pipe and then be transferred to any other pipe for delivery to a hub in the market area, where gas can arrive on any pipe and then be rerouted to any distribution company on different pipes.

The purpose of the hubs was to create the "missing links" that allowed gas to go from wherever it was available to wherever it was needed. Gas either can be delivered along a linear, balkanized, isolated pipeline geometry that was the historical precedent or can travel from hub to hub along any available pipeline. Instead of people in New York getting their gas off of Transco, Texas Eastern, or Tennessee, they could get their gas from a hub in Louisiana, a hub in the Rocky Mountains, a hub in Western Canada, and/or a hub in the Midwest.

Exactly how the gas gets from those hubs to the New York State hub is really not of any concern to the user. Hub operators could use the most efficient means of transportation available. If one pipe were full, gas could be routed on another pipe. For example, gas could be routed from Louisiana to Chicago and then to New York.

The problem with this approach is "stranded assets." Since hub-to-hub transport would be much more efficient, we would need fewer producer reserves and less pipeline capacity than we currently have. What do we do with all of the excess infrastructure? What about the rate-base companies that invested in these assets that are no longer needed? We've seen how the "gas bubble" has hurt producers. What happens when we combine this with a "transportation and storage bubble" that hurts pipelines?

Reengineering the gas industry

When you reengineer any industry, there are three critical elements: You have to reduce the capital investment per unit of goods or service delivered; you must reduce the cost of labor per unit; and you have to reduce the cost of capital.

In the gas industry, we must first reduce the amount of capital invested per MMBtu transported and burned. The test for every new investment should be whether it improves the capital efficiency compared to historical assets. When we at TPC invest a dollar, we expect it to be fifty times more productive than a dollar invested in simply expanding the old infrastructure—drilling more wells, looping more pipelines, or expanding the distribution piping system.

The second step in reengineering an industry is to reduce the labor costs per unit of goods or service. In this case, you have to reduce the labor cost per MMBtu of gas that is transported and burned. That most frequently means people will lose their jobs. In the gas industry, because it is so capital intensive and less labor intensive, there is a possibility that if we lowered the cost of gas at the burner tip by reengineering the industry, we might minimize any further dislocations. That would be my hope, but I'm not sure that it can be done, and it depends on how fast the market grows relative to how quickly it reengineers.

The third element for restructuring is to reduce the cost of capital. For the natural gas industry today, the crucial issue is sanctity of contract. Looking back in time at regulated monopolies, there was a concept called the "regulatory compact" that meant a regulated company turned over control of its business affairs to the regulatory entity. This regulatory entity told the company when and how much it could invest, and then set rates to recover that investment. In exchange for that kind of control, the regulatory body had an obligation to ensure a fair rate of return to the regulated company.

At the other end of the spectrum, a nonregulated industry such as computer manufacturing operates under the concepts of a market (that is, competition) and sanctity of contract. If two parties agree to a specific deal (for example, term, price, damages), then they can expect the other party to perform.

Regulator's roles

The real sticky point that we've got in the gas industry today is that we don't have either one of these situations. We don't have a fully regulated industry. And, in fact, as we "deregulate," a lot of regulators feel that they are no longer obligated to see that the company gets a fair return.

On the other hand, regulators still maintain control with a clause in all contracts that makes them invalid if the regulators change their minds. The regulators want to rely on market forces but retain the right to reverse contracts they don't like. I see that as fundamentally incompatible—not being obligated to see that the company makes money, wanting the company to compete in the market, but at the same time saying that the regulators can always have a recall on any agreements that are made. That raises the risk to both equity and debt capital and, therefore, raises the cost of capital.

To make market forces work, FERC [Federal Energy Regulatory Commission] has to allow sanctity of contract and must allow market rates anywhere there is competition. Customers must be allowed to buy gas under contracts that are determined in the marketplace. Once contracts are signed, they have to be sacred. You can't go back and plead for a regulator to change them, because that runs up the cost of capital.

Battle against technology

Regulators also have to think about what role they are going to allow technology to play. The concept of rate-base regulation is severely biased against new technology. When I came into the gas industry about fifteen years ago—I come from a high-tech background—I couldn't believe it. It was like stepping into "The Land That Time Forgot." The technology was basically late 1950s, early 1960s vintage as late as the early 1980s.

The reason for that is rate-base regulation. From the senior executives' viewpoint, there is no incentive to accomplish the tasks for less capital. In fact, if you have your choice of two projects, one of which costs $1 million and the other $10 million to accomplish the same task, you will rationally favor the higher one because you can get more dividend dollars for your shareholders.

Since one of the primary goals of technology is to reduce the capital cost to provide a given function, technology is anathema to executives in rate-base companies. If you're an executive in a utility, you don't want to see new technology, because that means that you can accomplish the same task for less capital investment.

If you're a regulator, you also don't want to see new technology. The last thing you want to have happen is for you to approve a $100 million project expenditure by a company that you're regulating, have them get half way into the project, be committed to spending the full hundred million, and then have some damn engineer come in and tell you, "Great news, I can do this same task and provide the same service for $10 million." As a regulator (frequently elected), you're causing the ratepayers to pay $100 million when technology can now do the job for $10 million.

So both the companies and the regulators are strongly biased against seeing new technology. When you open up competition, you begin to generate winners and losers. One of the things that distinguishes winners from losers is technology. All of a sudden we've unleashed a lot of technology that was generally available but was never used in the gas industry because of the regulatory bias against it. Regulators are going to have to decide how much technology they are really going to let in. More technology will produce more efficiency and lower costs at the burner tip, but more technology will also produce more "stranded assets" and reduce employment levels.

I think the third and final issue the regulators are going to have to look at is the question of monopoly practices and protection from antitrust behavior. When you have a regulated monopoly, it is exempt from most antitrust laws. As regulation "lightens" and regulators begin to use "market forces" and "competition," the regulated companies should begin to be treated like any other business.

The gas industry right now is only half way through the process of deregulation, and that's the problem. The companies are no longer fully regulated, but they are not able to just engage in free competition, either. As you start to lighten the hand of regulation, it's a simple question of, When can previously regulated companies be prosecuted for antitrust actions? Clearly, when they're fully regulated, they shouldn't be prosecuted for most antitrust actions. And clearly, if they are competing in a free market, they should have to abide by the same rules as Ford Motor Company and General Motors.

Competing with pipelines

The real question is, If I'm a producer, can I move my gas to any customer anywhere in the United States, or is the pipeline that I'm dealing with going to have control of me so that I can only move gas under the terms the pipeline tells me to, to the customers they tell me? And as a customer, am I going to be captive to a pipeline, able to buy gas only from that pipeline and only from producers that are hooked to that pipeline? That's the real issue. That's where competition comes in.

Ultimately, the competitor to the pipelines is going to be an organization of market hubs that act together as a system. There will be five, ten, or fifteen hubs that all exchange gas from hub to hub without regard to which pipe they use. These systems of hubs are the competition to the pipelines.

I'm less concerned about the consolidation of the pipelines and more concerned about being sure that we keep the playing field open for whole new paradigms, like hub-to-hub movement as opposed to linear movement along a single pipeline. Pipelines are now using tariffs to "capture" gas by extracting economic penalties for moving gas from one pipeline to another. If regulators allow this to happen, then we lose the economic efficiencies of allowing gas to go from wherever it is available to wherever it is needed. Regulators shouldn't let this backsliding occur.

Technological advances

Probably the most important technological change is the ability to create a market hub that can take gas in on any pipe and send it out on any pipe. The technology is not particularly sophisticated, but needs to be allowed to be developed. It must not be thwarted by regulatory rules or tariffs. This is what allows us to optimize the transportation grid instead of just optimizing individual pipelines.

The second technological change is salt storage. This is really important because it's what allows you to reduce the number of wells drilled and the reserve-to-production [R/P] ratio in the industry. Gas efficiency will increase due to a reduction in R/P ratios. Over the past few years, the industry has reduced the R/P ratio from around 16 to about 9. With this reduction, $140 billion of unnecessary inventory was removed from balance sheets. In the future, the R/P ratio will drop even further, down to 5 to 6 or even lower. This will allow the industry to reduce the capital held in nonproductive assets underground.

New technology is the key to lower R/P ratios. Improved finding and drilling technologies will allow companies to identify prospective reserves and use low-cost options to lock up the land. Drilling will be delayed until the gas is needed to meet market demand. When drilling is performed, new technology will ensure a much higher success level in achieving productive wells.

Remember, from a customer's viewpoint the important issue is not reserves but deliverability. When an LDC [local distribution company] experiences a peaking crisis, they really don't care how many Bcf of reserves you have in the ground. All they care about is how many Bcf of gas you can deliver to their city gate that day. Salt storage allows for the deliverability of the overall gas system to be greatly enhanced.

TPC's one little salt storage project that is in operation in Texas has a deliverability equal to half of all of Exxon U.S.A.'s production. You can turn it off and on with fifteen minutes' notice, and it is the equivalent of turning off and on half of all of Exxon U.S.A.'s gas production. We can only do that for a few days, but that will meet the demand during peaking crises. You don't have to have a large number of shut-in wells that are used at low load factor. That's a very inefficient use of capital.

It's much better to employ your capital in a salt storage facility where you can run your wells flat out twenty-four hours a day, 365 days a year, and meet the swings and loads by creating very cheap, flexible deliverability that matches the load, rather than turning the wells up and down. For that reason, salt technology is critically important.

The third technological change that we must see is the use of computerized information systems for title tracking of gas. I'm absolutely incredulous that we have an industry that changes about $80 billion a year on such an incredibly sloppy title chain and low level of accountability.

In the past, because of the government-mandated monopoly, if gas was in the pipeline, it only belonged to the pipeline. When it left the pipeline to the city gate, then it belonged to the LDC [local distribution company]. The pipeline had an obligation to be sure that there was always enough gas in the pipe, and any customer could take off as much gas as he wanted. That was the pipeline's "service obligation."

Now, the customer is only entitled to the gas he actually bought *and* that his supplier actually put in the pipe. We must start enforcing this. As the situation currently exists, customers don't have to pay for reliable service, because the title chain and custodial responsibility of the pipelines are so poor that they can count on "stealing" someone else's gas during a crisis. This means that buyers don't need to and won't pay a premium for reliability. But if buyers won't pay for reliability, then sellers can't and won't invest in reliability. Over time, the overall reliability of the entire system will degrade. *The* deadly thing that will kill the gas industry is failure to solve this title chain accountability problem.

Changing leadership

In the past, the biggest problems were regulatory issues. Companies were typically run by either regulatory attorneys or financial specialists in rates. They came to power in a company by being the best regulatory attorney or the best rate analyst who could get the highest rates at the rate hearings. Today, the problems are markets and competition, not regulatory.

The emerging chief executives are people who came out of fiercely competitive commodity industries. These newer executives are concerned about marginal costs, market share, and profitability, as opposed to rate-base and tariff considerations.

The personality of leadership is changing because the fundamental problem has changed. This new leadership is going to focus on the fundamentals: less capital invested per unit of gas transported and burned, less labor costs, and lower cost of capital. This will ultimately increase the market for gas and grow the industry. With this change of leadership, the "culture" of the industry is also changing. While competitors can be friends, the pleasant "collegial" environment will disappear.

I recently visited with a senior executive in the electric industry. After I related two or three anecdotal situations of how pipelines that were supposedly competing had seriously abused their remaining monopoly powers to the detriment of TPC, he was aghast and said, "Well, that's the gas industry. It comes out of the culture of the rough-and-tumble oil patch, where you guys go to the bar every night and get in fist fights and shoot each other. The electric industry is different. The management of the electric industry are gentlemen and they behave as gentlemen. I would never expect anything like that to occur as the electric industry deregulates, because there is a different class of individual."

I didn't challenge that, but I thought, just wait a little while. In a regulated environment, there are no winners and losers. Everybody comes out the same. You can have your nice little clubs and all the senior executives from the regulated entities can get together and discuss issues in a very positive way. If one takes more money out than the other one, it's not a zero sum game. The one who didn't get as much money needs to learn from the first one so that he can get more money out of his commission and rate payers.

When they become competitors in the market place, it will be a little different. There will be situations in which one company takes the money out, and then it's not there for the second one to take out. While competitors can be friends, the clubby atmosphere will change.

One of the clearest indications of change is a change in language. I'm a firm believer that language controls behavior. The big change in utilities occurs when they quit referring to their "rate payers" and start referring to their "customers." A rate payer is a passive sheep to be shorn. A customer is someone who has to be courted and won.

In the end, successful companies will emphasize the value of intellectual capital versus physical assets. The hiring, training, and retention of top-quality people will be much more important. Personnel will have greater value because there will be fewer but better people. The capabilities of the best people will be highly leveraged due to increased technology, and this allows greater investment in human capital. A key attribute of successful companies will be that they offer entrepreneurial positions in which key employees have an opportunity to share the wealth they create for shareholders.

CAROL FREEDENTHAL

principal of the Jofree Corporation

Carol Freedenthal is a principal of the Jofree Corporation, a Houston-based information services company that analyzes trends in the energy industry. He holds a B.S. degree in chemical engineering from Georgia Tech and has worked extensively in a variety of energy and chemical business areas, specializing in natural gas marketing, regulation, and engineering.

Before moving out on his own, Freedenthal was senior planning consultant for Mobil Oil's North American natural gas operations. Prior to joining Mobil he was manager of market planning and development with Superior Oil Company. He has also worked for Allied Corporation as director of commercial development, Kocide Chemical Company as general manager and chief operating officer, Kennecott Copper as plant manager, and Monsanto as a research engineer.

Freedenthal is well regarded throughout the gas industry, both for his optimistic nature and for his ability to stay on top of events, as his uncannily accurate forecasts of natural gas prices have shown. Jofree publishes a monthly *Update* newsletter that closely tracks the growing natural gas industry, and each January Freedenthal prepares an eagerly awaited price projection report that receives wide media coverage.

These days, if he is not traveling around the country as a consultant or visiting his first grandchild in California, Freedenthal is busy tracking the continued deregulation of the electrical power industry. His discourse is enlightening, because he is a visionary who sees opportunities and possibilities where others often cannot or will not.

*W*hen oil and natural gas started in the early days, everybody was looking for oil, and gas was a pain. If you found gas, you might cap the well and go on because what you wanted was the liquid. Nobody knew how to handle gas. There were only a few markets for gas.

In the 1950s, 1960s, and into today, we've seen a transition where people have drilled for gas only and want gas because it is an item of commerce by itself. It's no longer the weak sister to the oil industry. Now they both start out that you look for them the same way. The exploration is the same; the drilling and production is similar. But from that point on they really diverge and go separate ways. They come back again in the marketplace. But oil and gas are now competing for the same markets, not only in the United States and North America, but worldwide. On the world market we call it LNG—liquefied natural gas—because that's the way of shipping it. But what you're doing is competing with oil as a source of energy.

As motor fuel, gasoline developed from crude oil, and as a liquid is more efficient. But once you leave transportation, gas is much easier to use and cleaner. That's the beauty of it.

Around the world, we're really seeing gas grow at a much faster rate than oil. In the United States since 1985, gas consumption has increased an average of 3 percent a year. Oil consumption has gone down because we have to import oil. More than 50 percent of our oil right now is coming from outside of the U.S., whereas 90 percent of our gas is domestically produced.

So that's the transition we're seeing. And as we go longer into it, we're seeing the business solidify into a separate oil industry and a separate gas industry, once you pass the exploration and production.

The growth of gas has shown that the importance of oil as an energy source has diminished. In the 1972, 1978 OPEC declarations of their independence and their raising the price of oil, the Japanese were forced to go with OPEC. They needed OPEC because that was their major source of oil—their major fuel.

When we went into the Iraqi war a few years ago, the Japanese were on our side. The reason is that now 50 percent of Japan's fuel requirements are coming from natural gas, and a lot of that is coming from non-OPEC countries. Indonesia is a big supplier.

There's talk of building a pipeline from Russia through the ocean to deliver gas in Japan. A monstrous project—a 2,000–3,000-mile-long pipeline. It's technically feasible; it's a question of economics. But if it is done, Japan will have an alternative source of fuel to liquid fuels.

The independence or the more differences that you're going to have, the less anyone controls the system. So that's what you're watching happen. Gas is sharing in the world use of fuel and making it so that oil is not king any more.

The international oil companies fit in as they are really oil and gas companies, and the distribution systems that they have built to move oil are being turned to move LNG as well. It means that they are branching in their technology and in their capital expenditures to be prepared to go both ways.

All of the Americas

We usually talk of the U.S. gas business, but we ought to talk of the North American gas business, and if we really are sharp, we ought to be talking of the Western Hemisphere gas business, because the U.S. gas business is no longer just the U.S. You've got Canada with Trans-Canada, Nova, and Western Gas supplying gas into the U.S.

At the same time, you've got Mexico, which could be either a buyer or a supplier. That's the total North American picture. And through the pipeline system, I can have Canadian gas and be competitive selling that gas in the lower part of the U.S. In Florida, I can be competitive to gas that was produced in Texas. Economically, I can put gas anywhere I want essentially for the same price.

Let's go the next step. Now there is a movement in South America to develop an infrastructure. They could begin to bring gas from the south up or from the north down. It's just a question of economics and demand—how it fits into the development of the gas business. If it's too far to pump it in a pipe, we can build an LNG plant in Brazil and an LNG plant in Peru and bring it into the U.S. as liquefied natural gas. Or, build a pipeline and we could pipe it all the way up. You've got to get the demand and the supply in line. Otherwise you kill the economics.

In the transportation of gas, the electronic systems that we're bringing into control so we know where the gas is flowing are also adding to the ease of moving gas and lowering the cost.

Technology aids recovery

Across the board, technology is playing a tremendous role in oil and gas reserves that before were too expensive to develop. We now feel we can develop, because we can do it so much cheaper.

In this country, we will continue to drill because we need so much and we don't want to be totally dependent on outside sources. But the cost of drilling and the need to find oil in large pools is going to move it from being drilled by the majors to being drilled more by the independents, who can afford to find one well here and tender that well to get what it can produce.

There's a big potential in the U.S. that we have not totally worked over. In the normal delivery of oil, you only get about 30 percent to 35 percent of the oil in that reservoir. You can go to secondary recovery, where you might pump steam in or put CO_2 into the reservoir—something to extract more oil out of the ground.

We're also going to see movements to what we call tertiary recovery, where we can even do more. They may put chemicals down in the reserves to get the oil to come off of the rocks that it's bound to. That's going to produce additional oil. But we're not going to do much of that or we're not doing all that we could do now, because that adds to the cost of lifting the oil. There's so much oil available for cheap prices today that you can't afford that additional cost. But that could be the second stage in U.S. oil production when oil prices justify the additional cost.

We've improved on the technology of knowing where to drill. We used to use a two-dimensional picture to look where there was oil and gas in the ground. The marriage of the computer with the data developed in seismic determinations has allowed us to go to 3-D seismic, where we get a three-dimensional picture and can be much more accurate in where we drill a well.

Because of 3-D seismic, where before you might have had one dry hole in four or five drilling attempts and received zero value back for the millions you spent to drill the hole, now you might go ten or fifteen wells before you have a dry hole.

We've also learned to drill across the reservoir where before we drilled down into a reservoir—vertical drilling. We do a lot of horizontal drilling and that in itself gives you a much broader area that you can collect oil from. Drilling is an art. We've developed better ways of measuring where the hole is going and navigating it three, four, five miles into the ground. We also have better well-completion techniques.

Opening up the system

But let's be truthful about imports. If it costs me $12 to lift a barrel of oil in the United States and I can buy if for $12 from Saudi Arabia, where it only costs them $1 to lift it or maybe less, then the smartest thing you could do is use up all their cheap oil. Save our more expensive oil for the day that their cost goes up to $12, and then let's use my $12 oil because it's the same price.

It's not wrong to buy their cheap oil. Now, if they're getting $50 a barrel for their oil, then I ought to produce my $12 barrel to compete with it, because that's a nice margin. But if oil is only going for $14 a barrel, I ought to use all the cheap oil I can because I shouldn't use up my more expensive reserves.

The fact that you're paying money to them, and you worry about the balance of payment, really isn't something to get too concerned over, because we are getting a bargain for the money spent.

The other thing is bring the foreign suppliers into the country where they share their wealth back in our system. The more you can bring a commodity manufacturer to the consumer where he shares in the market risk, the more responsible he has to be. This has been done in the last few years.

We've opened up the U.S. refining and marketing capacity to OPEC countries, so the very suppliers who are part of the 51 percent of imported oil have a stake making them just as worried that if something happens on the selling side, they could get hurt just like we do. They're not going to let something happen. They've got to make sure that supply continues to flow, and as long as it's cheap, we ought to take advantage of it as much as we can.

We've done it with automobiles. You really don't know what's an American car anymore. A Honda is an America car. Depending on the model, it was made in Ohio. Or it may have been made in Japan. A Chrysler may have been made in Canada and is not an American car.

What you've done is you have assimilated and integrated the system so that if a foreigner were to try to play a game with you because he's got a lower labor rate or a lower raw material rate, he would hurt himself on the domestically produced cars if he tried to undercut the U.S. supplier.

I'm not an economist, but I think one can envision that we're really running to one world. If we mix the basic supplier with the final supplier so they share the same outcome, nobody will try to do something that could cripple the industry. That's a better way to go than an oil import fee or regulated system.

Smaller, better-trained

We are getting more economical in how we handle people and do things. The buzz word in the industry is reengineering. I think it's just a fancy way of saying that because we have computers, videotapes, and cellular telephone as ways of reaching and training people, we can cut down on the many levels of middle management needed to supervise the work force.

We've seen the oil and gas industry reduce itself roughly by a half million jobs over the last few years. Part of that is due to economics—we couldn't afford all the people. But the truth is we were an overpopulated industry. We had too many people doing what should have been one man's job. Part of that decline in numbers is that we've learned how to work more efficiently for the dollar spent. That's having a big effect on industry—not only oil and gas, but U.S. industry across the board.

What we need to lead this industry in the future are people who are adaptable. We still need the smart man. We still need the serious person who has perseverance and can follow through.

But what we really need, because things are changing so fast, is a creative person in that slot. And the creativity that he brings and the fearlessness of failure—you can't be afraid of failure in today's world, because the odds are that's the only way we can measure where you've gotten. It was once said "you haven't succeeded until you've failed."

We need that kind of leadership and we're seeing it. Ken Lay [chairman of Enron Corporation] is a good example of a fearless leader. He's not afraid of failure. And the younger people coming out of school who are doing well are ready to grapple with things they don't understand and don't know, but realize they're part of the business and have to be handled. And, to me, that's the change.

The big thing today is that the manager's got to have an open mind and be ready to change, because the world is crazy. I can't tell you why some things happen. They just happen, so you've got to be able to respond and that takes a creative person. Especially in the gas business, where we've gone from a very rigid commodity-oriented, regulated industry to a service-marketing industry with no regulation, you better be able to run with the best of them and be imaginative or you're going to left behind. And we're seeing that in the industry.

Continued efficiency

The industry is going to continue to restructure as we get more efficient and as new technology keeps changing things. Let's say you decontrol the electric industry and have a generating section, distribution section, transportation section; that distribution section is no different, except it's handling electricity, than the distribution group in natural gas.

Why not combine the two into one company? Go another step—you are now selling gas and electricity to the home. Let's sell them their burglar alarm—put in the alarm system. Let's sell them cable television. Let's sell all the home services that are delivered from one vendor. I don't care what the product line is he carries, let's send that all to the home.

So we could redistribute the system over different market lines than it is today. And it's unlimited—the potential ways we can go with these. If the electric automobile makes an inroad, that's going to develop a whole new substructure of how you're going to charge that car and keep up with the electricity.

Energy in itself has unlimited possibilities for being restructured because, in the end, you don't care whether you buy electric, gas, oil, nuclear power. All you want is the energy in your home to do what you need to do. The energy to drive your car, and it may come from new ways. All kinds of imaginative things out there.

Today the trend is to outsource and use your own people for direct core work. Suppliers of the new technology play a monstrous role in this industry changeover, because they are the agents of change. And in order for them to make their money, they are bringing you change. If you're going to go the status quo with them, they can't help you.

Risk-takers

These people are on the leading edge, and they do have failures in what they do. They come in with an innovative thing and it doesn't work, but that's part of the price of change. The sharper companies take that innovative risk, because everything in the world carries a risk. There's no risk-free system. We can measure risk and evaluate what's a safer thing to do.

But there's no place that you're totally risk-free. Innovation in itself carries a big risk. But the ones who take it and win do a phenomenal job and will be the leaders into the twenty-first century. Enron is a good example. There are some more out there that you can look at that are willing to take the risk and have done it and have done well.

There's a tendency—we see it within our own consulting group—to say, no, it won't work. Well, change your mind. Let it roam. On the electric car—my vision of an electric car—there are certain ways the electric car is wonderful. Got to get the price of electricity down. Got to get a better storage cell so we can store electricity in a car. We will need a new system of distribution of electricity.

Something like this might occur: You drive up to a parking lot. Your bumper hits the guard rail that is also an electrode and you recharge your battery. You've got a transistor in your car with your code number, just like in your cellular telephone. You get a bill at the end of the month from a company for electricity you used to charge your car—you don't have to do a thing.

You can bet Saudi Arabia goes to sleep worried—What if somebody thinks of a better fuel than crude oil? The whole existence of Saudi Arabia's economy is that they have one hundred to two hundred years of crude oil in reserve.

I worked on a shale project for Superior Oil [since taken over by Mobil]. We went back to management and said the project was doomed to failure. At the time we told them, and it was strictly economic, not technical, that Sheik Zaki Yamani, who was then the oil minister for Saudi Arabia, essentially said to the shale people, "You all are ridiculous chasing shale. I'll give you oil for your life for $32 a barrel."

He was going to do us a favor. Shale oil would have cost $80 a barrel to produce. We laughed and said, "He doesn't have $32 a barrel. Ask him would he do it for $18." Interestingly, we are getting Saudi oil now for $18 to $20 a barrel.

So you really need to have an open mind, and it's going to go across the board what you can do. And you don't want to be stuck like the Saudis, Kuwaitis, with nothing but oil.

Keeping up with change

I think the world has evolved into a fabulous techno-economic system. The emotional social skills are coming along slower. We're having more problems with those areas than we are with technology. The technology is too far ahead of the people right now. And we have a cultural lag that's causing depression and problems in the psychological sense that we've gotten so far ahead, people can't grasp some of the things we do.

If you stop and think that energy in the United States is roughly a $500 billion business, you've got the dollars there to really go. It's about a tenth of the national GNP. As long as you have that kind of money there, anybody who says energy is the wrong business—"I don't want to be in the energy business because it's an insecure area." If you say, "I don't want to be in the energy business because I'm in the hosiery business and I make more for each dollar invested," absolutely good reason. But if you just say it because one day energy is going to go, then you are wrong.

It's still an excellent investment. We've heard some bad stories in oil—the exploration and production business is taking the brunt of these low prices. But in the end, it's got to be a good business because there are certain things you can't live without. You can't live without food. You can't live without shelter. You can't live without fuel. Regardless of what you say, we're not going to quit driving; we're not going to quit heating our homes—what are you going to do? Freeze in the wintertime? And we're not going to quit making things.

But, you say, the price could go down so low. Well, the price can never go lower than what the natural resource deliverer has to have as a profitability to get it out of the ground. There may be short-term out-of-balances, but for the long term he's got to make enough to justify drilling and delivering it.

I talked with a big headhunter in the city. He's very good. He made millions putting people in the right companies. He called me up and said he had lunch with a president of an oil company and the guy told him, "Natural gas is a dead product. In five years there will be no natural gas. It's a secondary product that's an orphan to oil, oil is the major product. The only reason people sell gas is they don't know what to do with it."

I laughed. I said natural gas is becoming a product itself. People drill gas wells. They don't drill only oil wells now. We've based industries on gas. The ammonia industry is totally on gas. The methanol industry. It heats 56 percent of the homes. We're looking at automobiles and using gas. Good. Do a few of them, do what you can do economically. There are better ways. A liquid is easier to put in a tank than a gas. It would be nice if we could put solid pellets in there, it would make it even easier. Liquid is fine.

But gas is not going to go away. And anybody who thinks it is, is just blind. An $80 billion industry doesn't fold overnight. And that's just U.S. If we really talk in terms of North America, you have Canadian and you have Mexican and, like I say, you can go from there.

I think it's a thrilling time.

CONCLUSION

*T*he idea for this book came to me in 1974 while I was trapped in the middle of a mile-long line at the only open service station near my home in eastern Pennsylvania. It was during the Arab oil embargo. I never realized a country could be brought to its knees so easily.

Ten years later I moved to Houston, Texas, to work for a newspaper, and although I would not comprehend the full scope of what was happening for several years, I had a ringside seat that allowed me to witness the upheaval of one of America's greatest and most important industries: the petroleum industry.

I didn't give it much thought until 1989 when, in wake of the *Exxon Valdez* disaster, a tremendous public outcry was unleashed against the industry. A year later sentiment against the industry was still at a peak, as proven by the passage of the Oil Pollution Act of 1990, even as we prepared to go to war in the Persian Gulf for oil. To make matters worse, the industry was still reeling from the devastating blows inflicted by the depression of the mid 1980s.

After the war, in 1992, the Bush Administration, led by Secretary of Energy James Watkins, fought for passage of an energy bill that would provide some relief to the industry, particularly the hard-pressed independents and natural gas producers. As Watkins details in his chapter, the bill was the only piece of substantive legislation passed during Bush's last year in office. It came about despite fierce opposition from within the administration itself—not surprising, coming out of Washington D.C., where most economists and lawmakers usually endorse cheap energy prices over a strong domestic petroleum industry.

By this time, it was becoming clear to me, as an observer, that few people in this country cared where their oil and gas came from or how it was delivered to them. All they knew was that they wanted as much of it as they could get, and they wanted it as cheaply as possible. People took energy for granted.

The Pokorny report

Why does the petroleum industry have such a bad image when its products are so vital to our culture and the maintenance of our standard of living? In 1992, Gene Pokorny, chairman and founder of Cambridge Reports/Research International, offered an explanation in a study titled, "Great Expectations, American Views about Energy, the Environment & the Economy."[1] Pokorny's report offers an explanation as to why the industry is perceived in a negative light by the public and, subsequently, by their elected officials. It is worth reviewing here, for it helps clarify public attitudes and policy toward the industry. In brief, Pokorny says Americans link energy usage with their perception of environmental benefits and risks:

Fossil fuels in general are now seen by many Americans to be dangerous to use by society because of their perceived negative environmental side effects. . . . The perceived negative environmental effects of fossil fuels are two-fold. First, many Americans believe the usage of fossil fuels causes environmental damage (oil spills, acid rain, global climate effects). Second, many Americans believe their usage depletes a finite resource, and this fact is itself an environmental negative.

Research during a number of oil crises indicated that Americans held negative opinions about oil companies but liked oil. In the 1990s, Pokorny says, oil has become a four-letter. "They know they can't currently live without it; they know they're dependent upon it. Yet they worry about the environmental consequences of its usage and resent their very dependency upon it."

In 1990 and 1991, Pokorny's group conducted a series of research studies for the American Petroleum Institute that explored Americans'

1. This presentation was made at the Thirteenth Annual Arthur Andersen Oil & Gas Symposium held in Houston on December 16, 1992.

feelings about the oil and gas industry as well as ways the industry could alter and improve that perception. Findings revealed that

1. Consumers perceive little or no common ground with the oil and gas industry. More than other businesses, the oil and gas industry is seen as operating in an economic world dominated by an overarching commitment to short-term profits. Consumers believe this commitment to the short term precludes any commitment to the core values they cherish: human health and safety, family well-being, environmental quality.

2. Consumers resent their dependence on oil far more than their dependence on other goods and services such as the telephone and electricity. They feel the industry uses that dependence to "gouge" them. Thus, there are widespread feelings that the industry is not regulated as it should be, given its utility-like—from their point of view—character.

3. Consumers have a relatively simple image of the oil and gas industry, viewing it as a sort of middleman that merely exploits a natural resource. They believe the industry provides little "value-added" to a basic commodity that comes from the ground.

4. Opinion leaders do not display the same anger and bitterness toward oil and gas and the industry as consumers do. Their tone is more one of frustration that the industry fails to realize that "more is needed" to engender public trust and support.

5. As influentials see it, the "more that is needed" revolves around the tasks of improving the environmental performance of oil and gas and the industry, leading the country toward an energy future less dependent on oil in general and on foreign oil in particular, and building a new service relationship with consumers.

Pokorny's suggestions include working with the automobile industry to develop more efficient vehicles, developing alternative forms of energy and activities such as accepting and recycling motor oil and helping establish community recycling programs, and providing information on how prices are set and bringing together leaders to develop efficient, safe energy for the future.

The bottom line, Pokorny says, is that the oil and gas industry must convince Americans that it is truly committed to the core environmental values of American society.

The other side

I wanted to hear the other side of the story. Who were the people that ran this business that so profoundly impacts our lives? I wanted to know who they were and what they had to say about their industry, in their own words. I began to contact people I had often read about or who were referred to me by industry leaders. The industry has traditionally been rather tight-lipped, but to my surprise, nearly everyone contacted agreed to an interview. They were eager to tell their story and were extremely generous with their time and their opinions.

There was not a J. R. Ewing or Jed Clampett among them. They all struck me as hard-working people who had dedicated their lives to an industry of which they were intensely proud. They included the smallest of independents to the CEOs of our largest companies.

The thirty executives whose interviews are included in this volume represent a cross-section of the industry. They include mechanical and petroleum engineers, geologists, lawyers, bankers, a journalist, an environmentalist, and a philosopher.

They work in a business that requires huge amounts of investment capital to find, deliver, and refine products and to develop the technology to be able to work anywhere in the world, under the most austere conditions. It is also a business that forces one to make fast decisions worth millions of dollars and to have the willingness to take risks, sometimes very big risks. The payoffs can be tremendous; the losses can be just as enormous. One does not merely dig a hole in the ground and expect oil to come bubbling up—hence the title *The Oil Makers*.

Changes in the wind

As the project progressed, noticeable changes were occurring in the industry that changed the focus of this book. Instead of studying past mistakes, it became imperative to consider new opportunities for the future. The entire petroleum industry was in the midst of a wrenching restructuring that would permanently change the way it did business.

The industry had become too big as a result of the false expectation that oil prices would continue to rise. There was the issue of shareholder value to consider. Companies were downsizing, consolidating, or closing, eliminating hundreds of thousands of jobs.

A leaner and stronger industry was emerging, one determined to do whatever was necessary to stay in business. Technology was now driving the industry as innovations in 3-D seismic, horizontal drilling, and improved completion techniques were making it possible to explore, develop, and produce more oil and gas at lower prices. Technology also meant that fewer, but better-educated, people were needed.

But there was a specific purpose to this new application of technology: it would only be developed where there would be proven gains in efficiencies and where money could be made. No longer was technology considered a "monument" to progress.

Service companies were consolidating as well. A new era of partnering was beginning to unfold, as Halliburton Energy Services' Al Baker and the late Alton McCready of Schlumberger Well Services explain in intimate detail. McCready makes one of the most candid admissions of any executive, recalling how 1992 was the pivotal year for the industry when its leaders realized that to survive they had to take matters into their own hands and do whatever it took to effect a turnaround.

Drillers offered turnkey services as an alternative to traditional day work and assumed more of the risk on projects in expectations of higher rewards as producers, and service companies finally came to terms with each other's needs. Cooperation began to take the place of competition in terms of technology and service. Veteran oil writer Sam Fletcher, in reporting about major companies outsourcing for services that would eventually lead to partnering operations, wondered, "Why couldn't they have been doing this all along?"

Consolidation, cost-cutting, and new products and services were sweeping through the service industry. Nowhere was that better witnessed than at Weatherford International, which Philip Burguieres built into an oil-field giant by acquiring one company after another. New service companies that catered to a particular niche were also emerging, such as Les Mallory's Serengeti International, which helped

introduce the science of bioremediation into the cleanup of nonhazardous wastes.

The majors speak

The petroleum industry has traditionally looked toward the major
integrated companies for its leadership, and several of its most respected CEOs give detailed descriptions of its upstream and downstream activities.

A key issue in every interview is the continued denial of access
to what Phillips Petroleum's Wayne Allen terms "high-priority" exploration areas. This has made it imperative for the majors to turn their attention overseas in what has become a global industry. The United
States is considered a mature province, because more wells have been
drilled here in the process of looking for and producing oil and natural
gas than in the rest of the world combined. Add to that the regulatory
costs associated with drilling and producing and the occasional efforts
of low-cost producers to increase their market share by reducing the
price of oil, and you have a dramatic decline in domestic exploration,
Shell's Philip Carroll explains.

But as the majors have focused their attention elsewhere, they are
selling off many of their domestic properties to independent producers
who can operate them much more economically. For many independents, it has been a once-in-a-lifetime opportunity to acquire prime
properties. For the majors, too, it has been a chance to learn a few tricks
from the independents, Chevron's Ray Galvin says.

Carroll gives a deft account of difficulties in the downstream segment, where the industry faces costly government mandates relating to
the Clean Air Act for making reformulated gasoline while other alternative fuels are subsidized. As the public began to balk at the increased
costs, many metropolitan areas opted out of the program, despite the
fact that refiners had spent billions to retool to manufacture the
government-mandated product. And Allen notes that a growing number of foreign players who control the production of crude oil overseas
have entered the domestic refining and marketing business, making it
even more brutally competitive.

But technology seems to be the key to the future. It continues, the

CEOs say, to be the best ally of a cost-conscious industry's efforts to be competitive and stay heathy in a low-price environment. In particular, the development of 3-D seismic, which improves the clarity of subsalt images, has led to a resurgence of activity in the Gulf of Mexico, where Shell Oil is the acknowledged leader in deepwater exploration in what many call the last domestic frontier.

In a significant change, the CEOs discuss how stand-alone efforts are being replaced by strategic alliances within the industry and how, ultimately, technology has become a companion to downsizing. Improvements in computers and software now allow one geologist to be as productive as six or eight were in 1980, and the same can be said of engineers. A smaller worker force was inevitable, they say: good companies would have downsized anyway because technology allows one to do so without sacrificing productivity. Carroll, who directed a major rebuilding of Shell Oil's organization, also gives his view of the new relationship between company and employee.

Environmental principles also figure high on the agendas of industry leaders as they call for a system that would rely on more cost-benefit analysis before regulations are enacted that profoundly affect business. They, along with every other industry person I spoke to, were indignant about the allegation that they are unconcerned about environmental issues. Conoco's Constantine Nicandros discusses the establishment of its much-praised environmental program in 1990 in the aftermath of the *Exxon Valdez*. Galvin describes the ways in which Chevron employees are trained to shut down any operation that could cause an environmental incident.

The industry seems to share the public's concern for a clean environment as well as a concern for the safety of its employees who must work in those environments. Besides, it is good business to be environmentally aware, as they have learned—albeit sometimes at great expense.

Common themes

Some analysts say that it is a mistake to identify the "industry" as such because of the vast differences among its various sectors. Their interests often conflict, but during our conversations, several common themes emerged, including the fact that the very nature of their competitiveness precludes them from having an effective single spokesperson, despite the wishes of Secretary of Energy Hazel O'Leary.[2]

As might be expected, all voiced a strong distaste for government interference. Indeed, in their interviews, Roy Huffington and Hugh Liedtke, two of the nation's most prominent industrialists, found little positive to say about governmental efforts on the industry's behalf. They both desire a comprehensive policy that would take into account all of the nation's energy needs and resources, one that would also include nuclear power to provide some protection in an unpredictable world, they say.

The two oil tycoons confront the problems and opportunities that currently face independent producers, both domestically and internationally. Could a small independent company follow in their footsteps today? And Liedtke, who was still chairman of Pennzoil during our first meeting, provides a candid glimpse into the new era of corporate downsizing that has affected all of American business.

Everyone interviewed agreed to the need for regulatory standards—particularly for the environment—but complained that in too many cases those rules were unreasonable.

The operators

These industry leaders focus on their need for stable oil and gas prices, which would allow them easier access to loans and enable them to conduct operations in a more orderly fashion. How to guarantee that stability is another matter. Talk of a floor price, which was widely advocated by the independents and loathed by most of the majors, has faded, although it has reemerged since a 1994 report that the nation had passed the 50 percent dependence level on oil imports.

Foster describes how a new generation of management has had to learn to deal with price risks as the industry became market-driven starting in the early 1980s, when price controls were lifted for oil and then for natural gas. Today, a growing number of companies, with

---

2. O'Leary was the featured speaker at a luncheon of the Cambridge Energy Research Association in Houston in 1994. Her admonition to energy executives that they find a single spokesperson similar to the automobile industry's Lee Iacocca was met with dead silence.

Newfield a good example, rely on financial instruments such as hedging or swapping to better manage price risk. The transition to a competitive free-market system has been a difficult one but is resulting in a steady flow of supply at low prices for the consumer.

McCarthy, in particular, tells how the independents take a big hit from onerous regulations. He makes an eloquent plea for an end to the alternative minimum tax, which requires companies to pay a tax whether or not they make a profit. The independents also object to OPA-'90, which was passed as a result of the *Exxon Valdez* disaster and would raise the cost of insurance liability for anyone operating in marine areas from $35 million to $150 million, which would put companies like Newfield out of business.

Shipping executive John Ellis of Skaugen PetroTrans knows how expensive that law can be to transporters such as his lightering company. But he also takes an enlightened approach to OPA-'90, recalling how the business world overreacted to the Occupational Safety and Hazards Act. And he asks an important question: with imports expected to rise to 60 percent by year 2000 and many tankers around the world already aging, are not higher standards necessary? Like his peers, he does not oppose regulatory standards, but he prefers to be part of the rule-making process. (As this book goes to press, the industry, thanks in part to the lobbying efforts of Denise Bode, president of the Independent Petroleum Association of America, is making headway in several key areas, including the alternative minimum tax and OPA-'90. Environmental regulations based on a cost-benefit analysis also seem to be gaining favor as a new Congress with a conservative bent has begun to take a serious look at opening up the Arctic National Wildlife Refuge [ANWR] to drilling. What the industry is not likely to get is an end to the long-standing offshore drilling moratorium that prevents it from drilling almost anywhere except in certain portions of the Gulf of Mexico.)

Natural gas

Natural gas has always been regarded as the stepchild to oil, but the continuing deregulation of gas has led to a resurgence of that business and helped revitalize the entire petroleum industry. For the first time, production has been geared to that politically favored fuel instead of to oil.

Because it is largely a domestic business, the natural gas industry has led to the development of a competitive market system and may lead eventually to a total commoditization of fuels that could be easily interchangeable, as the barriers between all fuels are lifted, predicts Paul Anderson of Panhandle Eastern. The transformation of the natural gas industry from one that is strictly commodity-oriented and regulated to one based on service/marketing has also engendered a new style of doing business—one more imaginative and entrepreneurial.

Enron executive Kenneth Lay succinctly sums up the new era in the energy industry as we push ahead. As the international market moves toward more competition—following the path set by the United States—the entire industry will be market-driven, cost-driven, more agile, all the things that a competitive industry has to be that a regulated industry does not.

Summary

As we move toward the year 2000, this much is certain: oil imports will continue to rise because Americans love their cars but do not want any drilling in their backyard or in anyone else's for that matter. A tightly controlled market system should continue to make oil available for about $18 to $20 a barrel in 1994 dollars, unless OPEC decides to regain its market share or if a serious crisis temporarily disrupts world supply.

Politicians and economists say this country does not need a big domestic petroleum industry. Obviously it does not need one as big as the one during the ill-fated "boom" of the early 1980s, but a strong domestic industry is needed, nevertheless. Enlightened and courageous government leaders must ease their grip on producers and promote legislation that enables them to do what they have always done so well: produce oil and gas.

The world is not about to run out of oil anytime soon, but let us not forget that we fought a war for oil just five years ago.

Let's see where we are a few years from now.

JEFFREY SHARE
HOUSTON, TEXAS
SUMMER 1995

GLOSSARY

Alternative Minimum Tax—an additional tax system that was intended to keep businesses from eliminating their tax liability by denying them certain deductions. For the petroleum industry, this meant two key expenses could no longer be fully deducted: intangible drilling costs and the percentage depletion deductions.

Arctic National Wildlife Refuge (ANWR)—a refuge established in 1960 and originally covering 8.9 million acres in northeastern Alaska, bordering the Arctic Ocean. Congress addressed the status of the refuge in the Alaska National Interest Lands Conservation Act of 1980 (ANILCA), increased its size to 19.3 million acres, and designated it as ANWR. The act set aside more than 8 million acres as a "Wilderness" area and an additional 1.5 million acres as a "Coastal Plain" area for further study of oil, gas, fish, and wildlife resources. The U.S. Department of Energy estimates that ANWR may contain as much as 10 billion barrels of oil that could be recovered using conventional technology and more than 20 billion barrels that are obtainable through enhanced recovery techniques. Although a limited program of seismic work has been permitted on native lands, leasing or commercial development may not be undertaken in the Coastal Plain area unless expressly authorized by an act of Congress. Proponents claim that only 2,000 acres would be required for development. ANWR has become a battleground between the oil industry and the environmentalists, who feel that the area is the last frontier of wilderness and are determined to see it left untouched.

Austin Chalk—a geologic trend of nonpermeable but highly fractured limestone that stretches from South Texas north of Houston into East Texas and Louisiana. Natural gas is the driving force of development, which was made feasible by directional drilling.

American Petroleum Industry (API)—founded in 1920 and headquartered in Washington, D.C., this national oil trade organization is the leading standardizing body on oil-field drilling and producing equipment. The chairman, who is the head of a major integrated company, serves a two-year term and is generally considered the leading spokesperson for the industry.

Deepwater discovery—an offshore site located in at least 600 feet of water.

Derivatives—financial instruments, the returns of which are linked to, or derived from, the performance of underlying assets such as bonds, currencies, or commodities.

Directional drilling—deliberately deviating a well from the vertical at a controlled angle, typically 50–70 degrees, in order to bypass hard-rock formations or to reach a particular part of a reservoir. Horizontal drilling shares common technology, except that the angle of inclination is great, usually approaching 90 degrees.

Downstream—all operations taking place after crude oil is produced, such as transporting, refining, chemicals, and marketing.

Drill string (also referred to as drill pipe)—thirty-foot lengths of steel tubing screwed together to form a pipe connecting the drill bit to the drilling rig. The string rotates to drill the hole.

Enhanced recovery (EOR)—special field of activity designed to increase or prolong productivity of oil and gas fields beyond normal pumping and flowing operations. Secondary recovery techniques include maintaining or enhancing reservoir pressure by injecting water, gas, or other substances into the formation.

FERC Order 636—the Federal Energy Regulatory Commission's final step in completing the natural gas pipeline industry's transition to an open-access, unbundled environment where pipelines have little, if any, regulated sales business and operate primarily as regulated transportation-only pipelines.

Field—a geographical area under which one or more oil or gas reservoirs lie, all of them related to the same geological structure.

Fishing tools—instruments designed to recover equipment lost in the well.

Independent Petroleum Association of America (IPAA)—headquartered in Washington, D.C., this trade association represents about 5,500 oil and gas producers in thirty-three states and focuses on national policy issues. About one-third of the independents have between one and five employees, and two-thirds have twenty or less. These companies generally do not operate offshore in foreign countries. Their production is split almost equally between crude oil (55 percent) and natural gas (45 percent). They drill about 85 percent of all domestic wells, and an average of 61 percent of the wells they operate are strippers. Independents also produce about 39 percent of all U.S. crude oil (50 percent in the Lower 48 states) and roughly 66 percent of all natural gas. They find more than half of all new oil and gas reserves in the United States.

Hedge—method by which a purchaser or seller locks in a current fixed, minimum, or maximum price to protect against adverse price movements in the cash market.

Intangible drilling costs—expenditures, deductible for federal income tax purposes, incurred by an operator for labor, fuel, repairs, hauling, and supplies used in drilling and completing a well for production.

Interstate pipeline—transmission or carrying of oil across state lines, regulated by the federal government.

LNG—liquefied natural gas; achieved by cooling and compressing gas so that it turns into a liquid at minus 260 degrees Fahrenheit.

Lifting costs—the expense incurred in producing oil and natural gas.

Lightering—unloading cargo from large marine tankers into smaller tankers that are more able to enter shallow-water ports.

Logs—records made from data-gathering instruments lowered into the well-bore.

LOOP—the Louisiana Offshore Oil Port, located in the Gulf of Mexico about thirty-five miles west of New Orleans. Built in the early 1980s and owned by seven oil companies, LOOP consists of a terminal that can off-load three tankers at one time and pump the oil to onshore salt dome caverns, where it can then be piped to refiners. LOOP has a capacity of 1.2 million barrels per day, but generally off-loads about 600,000 to 700,000 barrels.

Margin—difference between sales prices and costs of production.

Marginal—refers to discoveries, wells, or fields that may or may not produce enough income to be worth developing at a given time.

MTBE—methyl tertiary butyl ether is a component of reformulated gasoline that, when used in the gasoline blend, produces an oxygenated fuel that emits fewer pollutants than most of today's motor gasolines.

Oil Pollution Act of 1990 (OPA '90)—passed by Congress in reaction to the 1989 *Exxon Valdez* accident, this act established new standards for oil spill response and liability. The most bitterly contested element was a move to hike from $35 million to $150 million the liability insurance requirement for drilling platforms and other offshore and coastal businesses handling oil products. Smaller independents claimed the requirement threatened their survival in the Gulf of Mexico. In mid 1995, Congress was expected to pass a bill restoring the requirement to the original $35 million threshold.

Outer Continental Shelf (OCS)—that portion of a continental land mass that constitutes the slope down to the ocean floor. The shelves are heavily sedimented, and it is believed they contain a large portion of the Earth's undiscovered oil and gas. The industry is primarily interested in exploring the OCS off Florida and California.

Primary recovery—the recovery of oil or gas from a reservoir by using the natural pressure in the reservoir to force the oil or gas out.

R/P ratio—reserve production ratio; a method of describing how many years it would take to use up the nation's proved reserves of oil and natural gas at current production levels.

Reservoir—a porous, permeable, sedimentary rock formation containing oil and/or gas enclosed or surrounded by layers of less-permeable or impervious rock.

Royalty—a share of the revenue from the sale of oil, gas, or other natural resources, paid to a landowner or the grantor of a license by the leaseholder.

Secondary recovery—a method of increasing oil recovery by restoring or injecting energy into a reservoir through the use of liquid or gas under pressure. One of the more common methods used is water flooding, in which water pumped through injection wells into the oil reservoir spreads out from the injection wells and moves toward the oil wells, driving reservoir oil ahead of it. The system continues until the fluid taken from the producing wells becomes nearly all water.

Severance tax—a tax levied only on the petroleum industry that is paid to a state by producers on each barrel of oil and each thousand cubic feet of gas removed from the ground.

Stripper—an oil well that yields ten or fewer barrels of oil per day, or a gas well that produces an average of less than 60,000 cubic feet per day. Nearly 75 percent of the oil wells in the nation are strippers, accounting for about 14 percent of oil production.

Subsalt—rock formations that may contain hydrocarbons lying beneath long horizontal layers of salt.

Subsea—wells in which the wellhead is not on the producing platform but rather on the seafloor.

Tertiary recovery—the third phase of recovering oil and gas, which usually involves sophisticated methods such as heating the reservoir to reduce the viscosity of the oil.

Three-dimensional seismic—three-dimensional images created by bouncing sound waves off underground rock formations; used by oil companies to determine the best places to drill for hydrocarbons.

Trans-Alaska Pipeline System (TAPS)—a crude oil pipeline that runs 800 miles from the North Slope of Alaska above Arctic Circle to the ice-free port of Valdez in southern Alaska. The joint-venture operation was assembled by eight participating companies. Construction began in 1974 and production from Prudhoe Bay, where oil was discovered in 1968, began in 1977.

Upstream—the oil and gas exploration and production activities of a company's operations.

Wellbore—the hole drilled by a bit.

Wildcatter—an operator who drills the first well in an area where no oil or gas production exists.

Windfall profits tax—a federal tax on crude oil production intended to tap excess industry profits anticipated after the removal of price controls began in 1979. The tax was phased out in 1988.

Workover—reentering a production well for cleaning, repairing, or replacing equipment.